Authors' Note

Where indicated with an asterisk (*), pseudonyms have been used.

To our dad.
You always believed in us.
Look at us now.

—Brad and Barry Klinge

Contents

	Prologue	1
1.	Growing Up Ghost-Obsessed	3
2.	The Phantom Regiment	14
3.	Tourists in Ghost Land	23
4.	Voices from Cell Ten	36
5.	To Catch a Ghost	51
6.	Chilling Evidence from an Ice Rink	69
7.	Echoes in an Abandoned Old Folks' Home	84
8.	Mysteries at Myrtles	98
9.	Ghosts of the Garden District	114

10.	Whispers in the Walls	125
11.	The Flying Tape Fiasco	135
12.	The Ghost in the Bedroom	144
13.	Ancient Artifacts and Disembodied Voices	158
14.	Library Specters	171
15.	Everyday Nutsacks and Other Disasters	181
16.	Sounds from Beyond	193
17.	School for Scares	203
18.	The Battle over *Ghost Lab*	212
	Ghost Hunting: The Basic Tool Kit	225
	Glossary	227
	Bibliography	235

If our personality survives, then it is strictly logical or scientific to assume that it retains memory, intellect, other faculties and knowledge that we acquire on this Earth. Therefore, if personality exists after what we call death, it is reasonable to conclude that those who leave the Earth would like to communicate with those they have left here. I am inclined to believe that our personality hereafter will be able to affect matter. If this reasoning be correct, then, if we can evolve an instrument so delicate as to be affected by our personality as it survives in the next life, such an instrument, when made available, ought to record something.

—Thomas Alva Edison (1847–1931)

The important thing is not to stop questioning.

—Albert Einstein (1879–1955)

Prologue

It was midnight and bitterly cold as we drove down the narrow weed-choked road that led to the abandoned nursing home on the outskirts of San Antonio. The few houses nearby were dark and quiet. Behind the long, low building designed in the shape of a cross, the grounds were unkempt and overgrown.

Most people would avoid this place even in daylight. Even standing fifteen feet away from the entrance, an ominous, foreboding feeling crept over you. No wonder the local cops were spooked by the hang-up calls they kept getting to 911. The senior care center had closed a year earlier, the power had been turned off, and the phone lines disconnected. The phone company was at a loss to explain the hang-ups. And when the officers on duty were forced to venture into the vacant building to "clear" the calls, they heard strange noises and saw shapes moving through the darkness.

We unlocked the doors and crept inside, shining our flashlights down the shadowy hallways. The temperature was hardly warmer than it was outdoors—it was cold enough to see your breath. There was something sinister, almost surreal, about the deserted facility. Half-empty glasses of moldy orange juice still stood on bedside tables. Wheelchairs sat empty along the halls. File drawers had been pulled open and medical records strewn over the floor. It looked as if the staff and patients had all rushed out in a panic and never bothered to go back.

We walked into one of the empty bedrooms lining the halls and switched on our digital audio and video recording equipment. Then we began asking questions aloud, making our presence known to see whether we could draw out any paranormal activity.

"Who are you? What is your name? Were you a patient here?"

Slowly, we became aware of a strange, unnatural tightness in our legs, as if someone or something was wringing our pant legs as hard as it could. It intensified, and a vague nausea began to envelop us. Were we imagining it or did the air in this room seem heavier and more oppressive than it had in the hallway?

"Man, my legs feel really tight."

"No shit? Because that's exactly what's happening to my legs right now!"

"Um . . . can we get out of here?"

As if in response to the question, the door suddenly slammed shut . . .

Growing Up Ghost-Obsessed

The toy store had been closed for hours—its doors locked, its countless Barbies, G.I. Joes, and plush animals staring impassively into the darkness. Suddenly, the whirrrrrr of wheels broke the silence as a skateboard flew off a shelf and careened across the floor, crashing into the base of a display. As the security camera footage rolled, a second skateboard leaped off its perch and sailed down an aisle, as if guided by an invisible joy rider.

BRAD: I've been fascinated by ghosts for as long as I can remember. When we were growing up in the 1980s, my older brother Barry and I never missed an episode of *That's Incredible* and the haunted Toys"R"Us I described above was our all-time favorite. But it left us lying awake late into the night trying to piece together what was really happening in that video. *How could viewers like us separate fact from fiction? Was the store really haunted, or was there some logical explanation for the bizarre activity? If it really was haunted, who—or what—was haunting it? And why?*

Where do you go for answers to questions like that when

you're a second grader? I tried the library at Rose Garden Elementary School in Universal City, Texas, a suburb of San Antonio. But when I asked for books about ghosts the librarian led me to a shelf filled with the Hardy Boys series, which was popular back then.

"That's not the kind of book I want," I told her in frustration. "Those are goofy kids' stories. They're made up. Don't you have anything about *real* ghosts?"

We checked the card catalog for nonfiction and found only one listing, a book simply titled *Ghosts*. It turned out to be a small paperback Scholastic Reader tucked back on an out-of-the-way shelf, collecting dust. While I was signing it out of the library, I scanned the short list of names penciled in above mine to find out who else had read the book. Sure enough, Barry's name was among them.

I started thumbing through *Ghosts* on my way back to the classroom and was instantly hooked. This was the real deal. The pages were filled with accounts of reported ghost sightings from around the world. Better yet, the book contained a number of grainy black-and-white photos of what the captions claimed were actual ghosts caught on film. These were classic pictures: a transparent shrouded female figure dubbed the Brown Lady of Raynham Hall appearing to float down a wide staircase in an old manor house. Choppy waters off the bow of the SS *Watertown*, where the faces of two drowned sailors appear to bob in the waves. I wouldn't learn until years later that the photos I was looking at were considered to be some of the most famous paranormal evidence of the early twentieth century.

When I showed *Ghosts* to Barry after school, his face lit up.

"Yeah," he said enthusiastically. "I remember that book! I liked it, too."

I checked out that book again and again, until I had memorized every page, every story about someone hearing rattling noises on a stormy night and glancing over his shoulder to find a spectral figure gliding toward him. I studied the photos obsessively. Could the faces be random patterns of light and dark that people misinterpreted as human features, just like when you sometimes discern a face in a tree trunk or a wallpaper pattern if you stare at it hard enough? I wasn't sure, but I loved contemplating the unknown.

Until I moved on to middle school and had to return *Ghosts*, the battered paperback spent more time with me than it did in the library. I finally convinced my parents to let me order my own copy. I still have it in a box tucked away in my house somewhere, though I don't have to rummage through my old school mementos to remember it clearly. I can picture the cover perfectly if I close my eyes. It featured a single ghostly figure in a hood and cloak that always reminded me of the Grim Reaper. Who knows? Maybe that image stuck in my subconscious mind and influenced our choice of logos for Everyday Paranormal years later. (The centerpiece of our company logo is a ghostly cloaked figure, with its hood obscuring its face.)

In addition to learning *Ghosts* by heart, Barry and I devoured every other nonfiction book we could find about the paranormal. We never found the subject scary—just fascinating. We read anything we could get our hands on that began with the premise that ghosts might be real and that examined the supernatural using the same matter-of-fact approach kids' science and nature books use to explain weather, volcanoes, fossils, or the insect world.

Unfortunately, there wasn't much of that type of reading material available in those days. This was before every home had cable TV and a computer, before the Internet became a household word. You couldn't just launch a Google search and locate fifty Web pages with detailed written accounts of hauntings, color photos, and audio files of disembodied voices. Tracking down the information we wanted was like finding the proverbial needle in the haystack.

Like most of the kids we knew, we went to church every Sunday morning. Naturally, what we learned there led us to question how ghosts fit into the picture: The adults in our lives told us that when you die, you go to heaven—or at least your soul does. So why would some people stay here as ghosts? Why wouldn't others? We started asking "what if" questions. We still ask them today. It's the cornerstone of our attitude and our approach.

I think it's sad that so many children tend to lose their sense of curiosity as they grow up; they get jaded. Not me. I'm still curious about everything. Even now I think like a little kid in terms of questioning what I see and hear. Sometimes it drives my wife, Jessica, nuts. We've got four children, ranging in age from three to thirteen, and I'll tell them, "Look at that star up there. Did you know that it could be light from a star that exploded three million years ago and the light's just reaching Earth now?"

"Can't you just let them enjoy looking at the stars?" Jessica sometimes wants to know.

But to me, asking questions and finding out the answers is what *makes* life enjoyable. And since the paranormal deals largely with the unknown, there's tremendous potential to ask "what if" and uncover groundbreaking new explanations. What could be more fascinating than that?

I tell my kids, "Hold on to your imagination!" All great scientists, authors, and innovators do. The polio vaccine would never have been created if Jonas Salk had lacked imagination—if he had just shrugged his shoulders and said, "Well, we've got to stick to the accepted way of doing things. We can't try anything new." If Albert Einstein had let detractors who challenged his work discourage him, the world would be completely different today.

Don't get me wrong. I'm not suggesting our work with Everyday Paranormal is on a par with developing the Theory of Relativity. We're just regular guys. But we do appreciate the importance of open-mindedness and radical thinking. Our quest for answers might eventually help to unlock a few of the secrets of paranormal activity that have mystified the world for so long. That's our goal, at least. And in the meantime, for two guys obsessed with ghosts all their lives, there's no better way to spend a Friday or Saturday night than holed up in a historic haunted landmark, whether it's a Civil War fort, a former speakeasy, or a mental ward.

BARRY: Looking back on our childhood, I've tried to figure out whether anything could have predisposed Brad and me for a career in ghost hunting. But aside from our fascination with the paranormal, we were your average football-obsessed Texas schoolboys. We lived in a residential neighborhood in Universal City, Texas, a suburb of San Antonio. We moved there amid the Bicentennial fervor of 1976 and stayed, as did our older brother and sister, until we graduated from high school. I left four years before Brad did, to join the navy, then returned to Texas to attend the

University of Texas at San Antonio, where I earned a bachelor's degree in kinesiology. Brad went off to college at Southwest Texas State University to earn a degree in business.

San Antonio was an idyllic setting for kids like us—plenty of sun and located a stone's throw from the Alamo, hallowed ground for Texans, especially those fascinated by history like we were. We were surrounded by evidence of our hometown's past—Spanish, French, and German heritage mixed with cowboy culture. It was evident in the architecture, the cultural events our parents took us to downtown, the subjects we studied in school. We always felt a strong connection to history, partly because we lived in a place steeped in it.

We didn't grow up in a moss-covered historic mansion or anything—just your basic ranch built in the mid-fifties. But, as you know if you've watched *Ghost Lab,* even nondescript modern buildings can set the stage for paranormal activity. The ground on which the building stands or the objects inside it might draw spirits there. A nearby power plant, a river, or even a thunderstorm might act as an inadvertent battery to fuel paranormal activity. And we did have a number of odd encounters in our house over the years that we suspected might be paranormal.

The first involved our maternal grandmother, who lived with us from the time we were born. She passed away when I was in the tenth grade. Very shortly after her death, I woke up to use the bathroom late one night. Hanging on the wall next to the sink was a plastic Dixie cup dispenser, the kind you often found in houses in those days. As I watched, the dispenser opened and the cup on the top of the stack slowly slid upward and outward until it separated from the cups below it, then fell to the floor. It wasn't until the next morning that I grasped how truly unusual this was. Had I been sleepwalking? It seemed unlikely because I remembered everything that had happened before and after the cup fell.

I spent a good twenty minutes trying to see if I could make it happen again. I shut the dispenser and waited for it to open. No luck. Then I tipped it open repeatedly to see what the cups would do. They either stayed put or ended up on a sharp enough angle that gravity took over and, because they were stacked so tightly, the whole sleeve of them toppled to the floor. You had to tug at them to get them to separate. So how had the cup detached itself from the stack and defied gravity the night before? Had one cup been loose and the movement of the dispenser somehow created pressure that pushed it upward? Or was it something more? Could our grandmother have been giving me a sign that she was still there in the house? My instincts told me she had.

There were other inexplicable occurrences too. More than once, I was in my bedroom when I glanced up at the doorway to see a person walk by. I would get up and look out into the hall, only to find that there was nobody there. For years we had an old-fashioned 1960s-style radio in the kitchen, and I would often see the dial turn, the needle flicking back and forth, when no one was touching it. Pictures fell off the walls on numerous occasions, too. Obviously there could be a lot of logical explanations for that. Flimsy hooks. Loose wires. Vibrations from movement in a nearby room or from construction on a neighboring street. A member of the family might have brushed past hurriedly and failed to notice that they had dislodged the picture. But we sometimes found the same pictures turned around so that the images were facing the wall. Every member of the household was adamant that they hadn't touched the pictures. So who had?

BRAD: Our dad died on Christmas Eve 2002, following a long battle with lung cancer. By that time, my siblings and I had moved away from home, but Barry and I stopped by whenever we could to visit Mom and help her around the house. Dad had spent his last months at home with an oxygen tank to help him breathe, and for a long time after he died you could walk into the house when there was no one there and still hear the bubbling and the steady *whoosh-whoosh* of the oxygen tank.

The strangest and most memorable occurrence was the grapefruit incident. Dad loved grapefruit. He would sit at the head of the table with a grapefruit half on a saucer and eat it with a spoon. One day several months after his death, I walked into the kitchen to find a fresh-cut grapefruit sitting on a saucer with a spoon next to it at the head of the table, right in front of the chair where Dad always sat.

"Uh, Mom, do you have any grapefruit in the house?" I asked.

"No," she said, puzzled as she followed me into the kitchen. "Why?"

We both stared at the grapefruit. There was no one else in the house. I checked with my siblings on the off chance that one of them had stopped by, cut up a grapefruit, and then decided to leave without eating it for some bizarre reason. They hadn't.

Did those early family experiences inspire our interest in the paranormal? Maybe. Skeptics would probably argue that all the reading Barry and I did about ghosts made us suggestible—it planted ideas in our heads that caused us to turn the ordinary into something extraordinary and to imagine that we were experiencing paranormal activity. But we don't think so. We have dealt with far too many clients who were lifelong cynics until they suddenly witnessed paranormal activity to buy into the notion that ghosts appear through wishful thinking.

BARRY: Does some sort of sensitivity to the spirit world run in the Klinge family? Maybe. But our research indicates that ghosts don't seek out receptive folks like psychics and mystics to contact. They appear to believers and nonbelievers alike. However, we are not the only Klinges who have had encounters with the paranormal. Our oldest brother, Paul, was sitting in a church parking lot near his home in Dallas one morning, reading the paper as he waited for his children to exit the church. He had stopped going to Sunday services several years earlier. Now, he was in the habit of dropping off his kids and picking them up every week, though he never set foot inside the building himself. He was engrossed in an article about the local news when he heard a noise nearby and glanced up to find a disheveled old couple—octogenarians and possibly vagrants, judging from their appearance—shuffling up to his car. They started asking him questions about the church.

"What do they do in there?" they wanted to know.

"Well, I can help you out," he said kindly. He got out of the car and walked them up to the church. They followed him inside and stayed close behind him until he found one of the church elders.

"These people would like to know more about your church," he told him.

The man gave him a blank look. "What people?" he asked.

Paul turned around and was shocked to discover that the old couple was nowhere in sight. He checked the pews and the parking lot, but there was no sign of them anywhere.

As a computer programmer, Paul focuses on numbers, data, and hard facts. He is probably the least suggestible person we know. And yet he swears to this day that the old man and woman were ghosts and that the incident was an intervention intended to lead him back into the church, in which he is still very active to this day.

Our mom used to claim our fascination with the paranormal drove her crazy. She was mortified whenever we talked about it in front of her friends, the neighbors, or even the extended family. "Are you two going ghosting again tonight?" she would ask. "Aren't you ever gonna give up that ghosting nonsense?"

"Ghosting?!" we would respond, cringing at the term. "Mom, that's not even a word!"

Like many people who are unfamiliar with the paranormal field, she found the idea of studying ghosts frightening. She also condemned it as disrespectful to God. She thought it was a topic no self-respecting Christian would want to explore. But paranormal study has nothing to do with devil worship. That's another field entirely. We have never bought into the notion that investigating ghosts runs counter to religion. Why should it? Both deal with the idea of existence extending beyond death.

Mom also seemed to be superstitious about what we were doing. She worried that if we focused so intensely on this stuff, it might start happening around us all the time. Believe me, if that were the case, and we could make ghosts appear on cue, then filming episodes would be a whole lot easier. Happily, after watching *Ghost Lab*, even Mom has changed her perspective on the paranormal. We hope that this book will do the same for other people, too.

2
The Phantom Regiment

BRAD: My second obsession as a kid was U.S. history. Even when I was little, I remember thinking, "Wow, I'm standing in a spot just twenty miles from the Alamo. What was happening here when the battle of the Alamo was being fought? Was this tree here?" If I tried hard, I could picture past times. I could almost envision myself stepping back in time. Perhaps that's part of what sparked my fascination with ghosts. Maybe they seemed like pieces of history or links to the past.

My family encouraged my interest and took me on a number of vacations to historic sites, from driving trips around Texas to letting me fly to Washington, D.C., on my own when I was fifteen to meet my dad on one of his business trips. He spent forty years with the Air Force and was one of the military's first computer programmers, work that often led him to the Pentagon. We had a great time together touring the White House, visiting the memorials, and taking a day trip to Mount Vernon.

I was never the kind of teenager who felt embarrassed to be seen in public with his parents. In fact, I spent a fair amount of time hanging out at home with them when I was in high school. I was athletic and I played football, but I was no party animal.

I stayed out of trouble. If my parents gave me a midnight curfew, I would be home by 10 p.m. I had learned years earlier that if you do what your parents want, they give you what you want. And by the time I was seventeen, what I really wanted was to take a historical tour of the original Thirteen Colonies.

So in 1990, just before I started my senior year of high school, my folks and I took a four-week summer vacation driving through the mid-Atlantic, stopping at points of historic interest like Independence Hall and Christ Church Burial Grounds in Philadelphia, where Benjamin Franklin is buried. We drove through Kentucky, then to Pittsburgh, where my dad had grown up and where my grandfather still lived. Our trip led us through Virginia—including Yorktown, Colonial Williamsburg, Jamestown, Manassas and other Civil War battlefields, and eventually to Florida, where we drove down to Key West and then ended with a visit to Barry—who was married and stationed at Homestead Joint Air Reserve Base—before heading back home to Texas. Along the way, we took guided tours, snapped photos of ourselves in front of historic highlights, and thoroughly enjoyed being tourists.

About halfway through our monthlong road trip we reached Gettysburg, Pennsylvania. It was early July, the anniversary of the famous turning point in America's Civil War. That stop on our route would prove to be a turning point in my future, too—and in Barry's. We drove into town late in the afternoon and found a hotel situated across the street from the Jennie Wade House. Mary Virginia "Jennie" Wade was the only civilian casualty of the famous Battle of Gettysburg. Just twenty years old at the time of her death, she was leaning over a bread trough kneading dough to make biscuits for starving Union soldiers early on the morning of July 3, 1863, when a stray Confederate miniball pierced two doors and struck her in the back, between the ribs of her whalebone corset. She died in an upstairs bedroom of the little clapboard house a few hours later. Wade is one of only two American women over whose graves the U.S. flag now flies 24-7 (the other is Betsy Ross) and her house is a popular tourist attraction, widely believed to be haunted. (Incidentally, we spent a night investigating the house in the summer of 2010 and came up with ample evidence to substantiate all those claims of paranormal activity, which we featured in the first episode of *Ghost Lab*, Season Two.)

Unfortunately, when my parents and I arrived that afternoon in 1990, the Wade House was about to close. I caught the last tour, but I didn't get a chance to videotape anything. I told the manager how disappointed I was and she said, "Why don't you come back in the morning? I'll let you go through on your own and you can film whatever you want."

I showed up early the next day and enjoyed wandering through the small historic brick building at my own pace, inspecting the bullet holes in the doors, navigating the slanting steps and low-ceilinged rooms. I was ending my self-guided tour in the basement when a green fog appeared at the far end of the room. As I watched, it moved across the cellar past me, dimming the lights as it went. Then it vanished as quickly as it had appeared.

"Huh," I muttered. "That was weird."

My parents were still at the hotel, but I didn't bother to tell them what had happened to me at the Wade House. After all, it might have been a trick of the light or some strange phenomenon caused by humidity. I now realize that what I witnessed was genuine ectoplasmic fog, a concentrated paranormal mist or mass visible to the eyes—not just in photos. It's the only time I have ever encountered it.

The next day was July 6th. We had finished breakfast and were heading out of Gettysburg and on our way to Philadelphia, planning to drive through York and then visit Amish country. My dad was behind the wheel, my mom was sitting in the backseat, and I was in the front passenger's seat, gazing out the window when I caught sight of a group of about twelve to fifteen men clad in Union soldiers' garb tromping through a wheat field about three hundred yards away. *Cool*, I thought. *They must be going to a reenactment.*

"Dad, can you pull over?" I asked. "I'd like to try and get those soldiers on video."

He agreed, so I grabbed my video camera, jumped out of the truck, and plunged into the waist-high wheat, trying to film as I ran.

Ahead, the men walked not in formation but ambled along in a ragged line. I could see them clearly in the bright sunlight—the vivid red, white, and blue of the Union flag rippling in the breeze above their heads and the deep navy of their regimental flag echoing its graceful movements. There were riflemen with guns over their shoulders and drummer boys, their drums slung casually across their backs and held there by worn leather straps.

The scene was so striking that I stopped and panned from right to left to capture it. There was the deserted wheat field, a line of gold below an expanse of blue sky, and in the midst of this pristine still life, a little knot of weary soldiers. If not for the modern technology in my hands, I might have thought I had stepped into a time warp and traveled back nearly one hundred thirty years to an actual moment from 1863.

The soldiers were nearing a copse of trees that flanked the field. I realized I would have to sprint or risk losing them in the woods, so I stopped panning and glanced down momentarily to switch off the camera. When I looked up again, the men had vanished.

"Where the hell did they go?" I said, confused.

I raced ahead and reached what I realized was little more than a line of trees. I squeezed between the trunks and emerged breathless on the other side, expecting to find a field full of reenactors and spectators. To my amazement, I saw no people, no cars, and no regiment of ragtag Union soldiers—just a grassy hillside rolling down to a shallow valley, which sloped back upward in the distance.

I spun back around and scanned the field wildly. I was the only one out here. There was not a sound aside from the wind rustling the leaves and the endless rows of wheat. There were no men's voices carried on the breeze, no sounds of movement, not even a car passing on the highway. The silence seemed suddenly eerie and overpowering. How could a dozen men disappear? There was no place in the field, the trees, or the valley on the other side for them to hide.

And then it hit me: I had just seen a ghost regiment. After years of reading about ghosts and wondering if they really existed, there was no longer any doubt in my mind. They did.

Was it possible that I had managed to capture them on video? Shaken, I hurried back to the car where Mom and Dad were waiting, rewound the tape, and pressed Play. Sure enough, there were the soldiers just as I had seen them moments before. Was this physical evidence that ghosts existed? It certainly seemed to be.

I knew my mom would be unnerved by what had just happened and there was no way to tell Dad in private, so I kept the story of the vanishing regiment to myself until we reached Barry's place in Florida. "You gotta see this!" I told him the second my parents were out of earshot. He was as floored by the footage as I had been by the actual event. We watched the tape over and over. There were the bright colors—the azure sky, the bold reds and deep blues of the flags, the green of the trees behind the men—and the vivid detail in the uniforms, the drums, and the rifles.

"But look at the faces," Barry remarked quietly. "Everything has detail except for the faces. There's no skin tone. They're like shadows."

One of the questions people often ask me when they see this videotape is, "Why did you have the camera off when they disappeared?!" Because I assumed I was looking at living, breathing human beings! I expected to begin filming in earnest when I reached a full reenactment just beyond the trees. If *you* spied a group of soldiers walking through a wheat field, would *you* expect them to disappear? Even today, as a paranormal investigator, if I were walking through that field and spotted men in Civil War garb, I would assume they were flesh and blood. Countless clients of ours have echoed that same sentiment. "I had no idea I was seeing a ghost," they tell us, "until he vanished a few seconds later."

BRAD: I had one other memorable experience in my early years. Whether or not it was paranormal, you be the judge. Here's how it unfolded: One afternoon about a week before Halloween during my freshman year at Southwest Texas State University, I was hanging out in my dorm room attempting to study, but with little success. Outside the windows, the sun was shining. Inside, it was noisy and chaotic because my room was full of guys from my floor, all crowded around one of the desks, talking and laughing. A kid who lived down the hall had found a Ouija board somewhere and dragged it out in honor of Halloween. Everybody was hovering over it, asking it questions.

I didn't put much stock in those sinister warnings about the boards opening portals to the spirit world. How could a product mass-produced by Parker Brothers do that? Still, the whole thing smacked of witchcraft and black magic, and those were offshoots of paranormal study that Barry and I both disliked.

"Brad, you wanna try this?" one of the guys asked me.

"Nah, you go ahead," I told him. I glanced absently toward the window, wondering whether I should go out for some fresh air or head over to the library to study. Next to the window stood a towering stack of road signs that my light-fingered roommate had been collecting after dark from every construction site and roadside he could find. It was one of his favorite hobbies.

As I watched, the topmost sign in the stack moved slightly. *That's strange*, I thought. Those signs were massive, so heavy they were hard to lift. And they had been stacked there without moving for weeks. Even loud music and slamming doors didn't budge them.

Maybe I had imagined it. A few seconds later, I glanced over again. The octagonal sheet of red and silver metal on the top was *definitely* scooting slowly but steadily across the signs below it.

"Guys, you don't want to do that," I warned.

Not surprisingly, they ignored me.

"I'm serious. Y'all better stop messing with that thing."

"What's the matter? You scared?" one of them taunted.

No sooner had the words left his mouth than the stop sign flew across the room on a downward arc and caught him hard on the shin, leaving an inch-long gash in the flesh above his sock. He let out a yowl of pain.

"Holy shit!" the guy next to him cried, realizing what had happened.

It was pandemonium. All bravado forgotten, a dozen panicked teenage guys bolted for the door at once, me right along with everyone else. We nearly trampled each other in our haste to get out that door.

Everyone laughed it off later, though they still conceded that what the sign had done was weird. I kept trying to work it out in my mind, examining the signs more closely when I ventured back into my room later. Was it possible the vibrations from so many large bodies in a small space had jarred the sign loose? Had it been precariously perched all along, ready to fall? Did heat from the sun streaming through the window cause the metal to expand and shift? Or did they actually open a portal for playful or menacing spirits, drawn by all that human excitement and energy in the room? Today, Barry and I would use technology to gauge the likelihood of the sign moving. We would attempt to replicate the scenario with a dozen large men with booming voices gathered in the same room, making noise and moving around to see if we could get the sign to repeat its unusual action. But this was years before Everyday Paranormal; it didn't occur to me to try and replicate the incident or use technology to see if I could uncover the real root of what had happened. I just resolved that, whatever the cause, one thing was certain: It would be the last time anyone brought a Ouija board into my college dorm room.

Tourists in Ghost Land

BRAD: Before Barry and I became paranormal investigators—back when the term was still unknown—I channeled my fascination with the supernatural into taking ghost tours. They were one of my favorite ways to spend a Friday night. I've lost track of the number of lantern-toting guides in old-fashioned clothing that I've followed down cobblestone streets, into overgrown cemeteries, old churches, back alleys, and former speakeasies all over the United States. I took my first ghost tour when I was in college. I had a weekend job driving a Zamboni for a local ice rink called the Crystal Ice Palace, and I often rounded up my friends from work to tag along with me on tours in and around San Antonio. One of those friends was Steve Harris, who shared my fascination with the paranormal and who, as you'll know already if you're a *Ghost Lab* fan, is an original member of our crew at Everyday Paranormal. I call him the friendly skeptic. He's the life of the party—always amiable, always smiling, and *nothing* bothers him. He's incredibly tech-savvy, which is a definite plus on our team, but even after all he's seen with Everyday Paranormal, he's managed to hang on to a healthy dose of cynicism about the existence of ghosts. A ghost could probably slap him in the face and he'd still find a way to be skeptical.

Steve wasn't the only friend I lured in. After I graduated from college with a degree in business, I worked as a Human Resources Information Systems (HRIS) Implementation Consultant for a number of national companies. I struck up a friendship with two of my coworkers, Hector Cisneros and Jason Worden, and it wasn't long before I was dragging them along on weekend ghost tours, too. Like Steve, Hector and Jason caught the ghost-hunting bug and became founding members of Everyday Paranormal.

Hector soon became our very own Inspector Gadget. He was always showing up with some crazy new data logger. Still does. He worked for a medical equipment company and had a lot of contacts in the electronics industry, so he started supplying us with equipment, from walkie-talkies to EMF detectors and sometimes random gizmos he found a way to weave into our investigations. He was 100 percent reliable—if he said he'd show up, he did. If he promised to bring a piece of equipment, he never forgot it or brought one that was missing a battery—and that counted for a lot in our early investigations. He's probably the quietest, most analytical member of the team. That might be his basic personality, the fact that English is his second language, or simply that the rest of us talk *a lot* and it's hard for him to get a word in.

Jason gave us yet another dependable supporting player. Like Hector, he could be counted on to come through every time. He's loyal, conscientious, hardworking—everything Barry and I could want in a team member. He was also brimming with enthusiasm. When we started talking about forming a team, he was the first one at the table.

My job involved a lot of travel back then and no matter where I went on business, I made a point to include a ghost tour. My routine was to check into my hotel, drop my luggage, and then—in whatever free minutes I had before meeting a client or heading to the local office—start scouting around for ghost tours in the area. I would scan the Yellow Pages, skim the newspapers, ask the concierge and front-desk staff, or call the local visitors' bureau for help. Nine times out of ten, I would find one I could join while I was in town. I've taken ghost tours in Chicago; New Orleans; Baton Rouge; Colonial Williamsburg; San Francisco; Hampton, Virginia; Kissimmee, Florida; Omaha, Nebraska; several cities in Canada; and dozens of other places.

The tours were almost always in the evenings, which meant I could put in a full day's work, maybe even have dinner with colleagues, and still catch a dusk or late-night tour. Some of the tours I took were pretty goofy, but I found the majority fun and interesting, and they provided a great way to indulge my hobby. Borrowing the analytical approach I remembered from my favorite childhood book, *Ghosts*, I started to take notes and amass some informal data about the paranormal based on what I learned on the tours.

For example, virtually every town in America offered at least one or two weekly ghost tours. Was it just a gimmick for the tourists? If so, at the very least this suggested that a huge portion of the traveling public in the United States shared my interest in ghosts. Assuming the local tales about apparitions were based at least to some degree in fact, it suggested to me that paranormal activity was pretty common and that it could occur almost anywhere—from deserts to swamps to chilly New England ski towns. Ghost stories took place in every season and every setting. Remote farm towns boasted tales of local hauntings just like huge, sprawling cities did. It didn't surprise me that historic cities would be steeped in ghost lore, but so were modern ones. Ghosts seemed to be an intrinsic part of the lore and culture no matter where you went in the United States. I had read enough to know that you could make the same statement about virtually every culture around the world as well as about ancient civilizations, including the Greeks and the Romans.

Judging from all the stories I heard on my tours, manifestations were most common at night, though some occurred in broad daylight. Why, I wondered, would night be preferable for ghosts? Is it quieter then? Are we just more apt to notice ghosts at night? Maybe people were distracted by background noise and activity during the daylight hours, so they simply failed to hear or see the ghosts around them. Also, the bulk of ghost sightings happened to people when they were alone. Why? Was this a coincidence or something more?

What types of buildings and locations did ghost tours most often include? What kinds of apparitions did the guides describe repeatedly? Were there any patterns that might have scientific significance? Were dark and stormy nights truly more conducive to apparitions? If so, why? And why were there so many reports of women in white? Why were ghosts so often seen wearing Victorian clothes? The list of questions kept growing, though it would take years to start piecing answers together.

One observation really resonated with me—and it still does: Judging from the stories the guides told, most of the people who witnessed paranormal phenomena were just ordinary folks. They were schoolteachers, bus drivers, doctors, construction workers—people in all lines of work, all ages, all ethnicities, men, women, and children. A lot of them didn't even believe in ghosts until they had a paranormal encounter. That meant you didn't have to be looking for a ghost to find one. You didn't have to be a "sensitive." In fact, my research suggested that the vast majority of people who had paranormal encounters weren't expecting to see a ghost. Most of the time they hadn't *wanted* to see one and hoped it would never happen again. So, why did the ghost appear to them? Were they just in the right place at the right time?

And what of the ghosts themselves? Was there some common denominator in how they appeared? Did people see them straight on or were they always glimpsed through the corner of the eye or through a window? Did they appear in a gassy transparent form or were they solid enough that the witness mistook the apparition for a living human? I heard stories about both types of apparition on my tours. Why would there be different types of ghosts, some misty and some able to simulate flesh and blood? Why would some actually appear and others manifest only as disembodied voices or slamming doors? Gradually, I gathered enough little bits and pieces to start formulating some of my own theories about how all this might work. I'll tell you more about them in this book.

I would often pull the guides aside to ask additional questions and find out where they had learned so many of the area's ghost stories. A lot of them were historians or at least history buffs. They knew a great deal about local characters, architecture, past events, and the customs and superstitions of the people who had settled that area years earlier. But when it came to information about the actual ghosts, they were as much in the dark as I was. They were just enthusiasts like I was, intrigued by the paranormal and its role in history.

BARRY: Like Brad, I found ghost tours fascinating. When I was in the navy stationed in Florida, my wife, Kim, and I took ghost tours and day trips to visit famous haunted locations along the eastern part of Florida. I read books and watched movies about ghosts just like I had when I was a kid, and I paid close attention to anything I read in the news about the supernatural. It was still a huge interest of mine.

When I got out of the navy and came back to Texas, I started a career as a paramedic while taking college courses, eventually earning my degree and becoming a special education teacher. I made between 5,000 and 6,000 EMS calls and, believe me, I saw a lot of dead people. You can't do that kind of work without giving some serious thought to death and the possibility of life after it. Were any of the people I encountered in my work likely to become ghosts? If so, why? Would those who died in traumatic ways be more prone to it? How long after death would it take for a ghost to start manifesting? What about all the people I saw who had near-death experiences—who were technically dead for a few seconds but whom we resuscitated? Did they brush against the spirit world? And if they did, would it affect them? How? Brad and I batted around a lot of theories, but we still had more questions than answers.

In addition to walking tours, Brad and I took several local bus tours together in those years with Docia Williams, a local author and historian with a flair for storytelling. One of our favorite fixtures on the San Antonio paranormal scene, Docia ran a Spirits of San Antonio event every October where she loaded charter buses with enthusiasts like us and took us to visit one of the city's most haunted restaurants (the Alamo Street Restaurant), the old historic King William District, and a number of houses said to be hotbeds of paranormal activity. Along the way, she told us ghost stories and interesting historical tidbits, and gave us a chance to explore each of the places on our own. Brad and I made a point of looking for logical explanations in each building we toured. Was there a mirror that might have thrown a reflection onto a wall where people thought they saw an apparition? Was there an unusual pattern in the woodwork that somebody might mistake for a human face? We thought we were just having fun, but in hindsight we were laying the groundwork for the investigative methods we would develop in the years to come.

During a few of these tours, we had our own mild paranormal experiences. On one occasion, a flashlight in a rechargeable holder attached to the wall shot up in the air just as Brad and his wife walked past it. On another, a chandelier started swinging. The most dramatic incident happened during a Docia Williams trip we took with our wives and some of Brad's friends from the ice rink.

We were visiting a place called the Lavender House, which had a sinister reputation for being haunted by the ghost of an elderly woman, a former owner whose favorite color had given rise to the building's name. According to legend, the woman disliked men and resented their presence in her house. After her death, her property was sold to a succession of young couples. In each case, the husband died unexpectedly shortly after taking up residence there. Even stranger than that was the fact that whenever anyone attempted to repaint, the walls would inexplicably revert to lavender. People would buy gallons of white paint at local hardware stores, check the color before leaving, then get back to the house and open the paint cans. To their amazement, they would find lavender paint inside them. There were also reports of the house's antique grandfather clock ending up on the floor repeatedly with the internal workings pulled out, as if to stop time.

We had just climbed the staircase and started to explore the second floor of the Lavender House when I began to feel nauseous. I thought maybe I was coming down with the flu. I glanced over at Brad and noticed that he looked a little green around the gills, too, so I pulled him aside and asked if he was all right.

"I don't know, man," he mumbled. "I'm not feeling so good all of a sudden."

"Me, too," I told him confidentially. I was getting increasingly lightheaded and short of breath. I tried to remember if Brad and I had ordered the same meal at dinner a few hours earlier. Could it be food poisoning?

"Are you feeling okay?" I asked my wife.

"I'm fine," she responded, surprised. "Why?"

"Because I feel like I'm about to pass out."

I leaned against a wall to steady myself. Could there be some legitimacy to the stories Docia had told us about the ghost that lived here? "Are any of y'all feeling queasy?" I asked the rest of our group. Sure enough, every one of the guys confessed to feeling sick. The women all said they felt perfectly normal.

"I think it's this house," I said. "Can we get out of here . . . now?"

Back out on the street, we took deep grateful gulps of fresh air. Within minutes, the shaky, nauseous sensation vanished for all of us.

If we were to investigate today, Everyday Paranormal would do some research to make sure there weren't any chemicals or environmental factors in the building that might trigger nausea. Maybe there was something noxious in the roof, which would explain why the tallest people in our group felt its effects first. If we came up empty-handed, we would seek help from an expert specializing in household pollutants like black mold and other toxins. We might have paint chips or carpet samples analyzed at a lab. If we found no logical explanation for the physical symptoms the house seemed to be generating, we would launch an investigation and attempt to determine whether the source might be paranormal. But I'm not sure we could convince our original gang to take part. Most of us still feel a slight wave of nausea whenever we walk into a room painted lavender—and not just because it's bad taste in wall color.

BRAD: Another early route I took to expand my knowledge about ghosts was to attend paranormal conferences whenever they were held in San Antonio, usually once or twice a year. The one I remember best took place in 2002, at the historic Menger, a famously haunted hotel built in 1859 next door to the Alamo, and included an entire Saturday of lectures by paranormal authorities from all over the region. I started the day full of anticipation, but with each lecture my energy level plummeted. A few speakers presented photographs, but the vast majority were men and women who claimed to have special talents for communing with the dead. They told mystical tales of having "just known" or "gotten a strong sense" that there was a ghost sitting in the corner whose name was, say, Hannah and that she was from 1887.

I almost got up and stormed out. I didn't want to hear this; I didn't buy any of it. How could anyone believe these stories when there was absolutely no way to verify them—no physical supporting evidence to show the audience? I wanted to see the ghost with my own eyes. I wanted to hear it and look at pictures of it. Why wasn't anyone examining the topic of ghosts logically? Why weren't they offering any theories about why apparitions manifest in the first place and what allows us to see or hear them?

In one portion of the conference, participants were allowed to wander through the halls of the Menger and take pictures. It was the perfect setting for ghost hunting. Over its long and illustrious history, the hotel had hosted such famous names as Teddy Roosevelt, Robert E. Lee, Ulysses S. Grant, Mae West, and Babe Ruth. Its official literature claimed that no fewer than thirty-two different ghosts had been seen here and that it wasn't uncommon for employees to see kitchen utensils moving of their own accord, possibly guided by the invisible hands of former cooks.

Maybe if I was lucky, I would spot the famous phantom maid Sallie White, a onetime employee who was murdered by her husband and buried at the hotel's expense. White's ghost supposedly wandered the corridors clad in a long gray skirt and a bandana, carrying an armload of towels.

I was walking along an upper floor when I rounded a corner and felt as if I had suddenly stepped into a refrigerator. The air here was ice cold. I glanced down the hall and to my surprise I saw two young boys standing at the opposite end. One looked to be about ten and the other was probably seven. What struck me as odd was their clothes. They were wearing old-fashioned knickers and flat caps, like the kids who hawked newspapers on corners in the 1920s. Weird. *Maybe someone is shooting a movie at the Menger,* I thought . . . and in the split second it took this idea to flash through my mind, they vanished. They were there one moment, and then they were gone. I blinked and rubbed my eyes. Was all this talk about seeing ghosts getting to me? Was this sheer power of suggestion? After a day of feeling annoyed at people who asked me to take their word about seeing apparitions, I wasn't going to be a hypocrite. I whipped out my camera and started taking pictures of the hall. As I did so, several other conference attendees strolled around the corner behind me. "Oh, man, it's freezing here!" one exclaimed.

Later, I pulled one of the hotel employees aside and asked her if there had been any reports about ghosts on the floor where I had seen the boys. "Actually, they do report seeing two children there," she said. "People think they're ghosts of guests who were here in the early 1900s."

I asked myself a lot of questions after that. Why did I see those boys that afternoon? Do I have some sort of special power? No. I decided that I must have happened to look in the right direction at the right moment. Was the cold spot related to the ghosts or was it coincidental? Maybe the air-conditioning was just particularly concentrated in that area. Nowadays, Barry and I would take baseline readings to gauge the average temperature in the corridor. We would use data loggers and handheld digital thermometers to alert us to any dramatic drops or spikes in the temperature. We would return the next night to find out whether the sudden change might be a regular occurrence triggered by the building's central cooling system. We would note whether any other paranormal data correlated to the temperature change—a ghost sighting like the one I had at the Menger, for example, or a disembodied voice caught on tape. We might also do a linear sweep of the corridor, which means placing a straight line of electronic data loggers along it to take readings of temperature, humidity, and EMF (electromagnetic field) level every two seconds. If an entity moved through the hallway in a given time period, this would allow us to track its movement.

The cold spot in the Menger got me thinking about energy. Had energy in the form of heat drained out of the environment when the ghosts appeared? If so, had they repelled it, absorbed it, or maybe channeled it into the ability to appear in a physical form I could see momentarily? Barry and I kept formulating ideas about the ways in which ghosts appeared and trying to figure out how to differentiate between genuine paranormal activity and real-world phenomena that might be misinterpreted as paranormal, though we had no idea that other people were thinking along the same lines.

BARRY: The idea of ghosts is ancient, but the science of paranormal investigation was in its infancy a decade ago. When fact-oriented ghost research suddenly started to catch on, Brad and I were floored. So, we weren't the only ones asking these questions after all! We watched every documentary and every episode of TV series like *Ghost Hunters* on the Syfy channel. At last, people were starting to take a realistic approach to the paranormal world. After every show we saw, Brad and I talked about what we thought was being done right and what was still lacking. We respected the other teams out there, but we realized our ideas diverged in many ways. We craved more data correlation, more out-of-the-box thinking. We didn't just want to hear, "Yes, it's happening." We wanted to know *why* and *how* it was happening.

We used to talk about what *we* might be able to bring to the table if we had our own TV show, but we never considered it a realistic possibility. In fact, as you'll find out in this book, we never set out to star in our own series—TV found us. We're not complaining. We've loved filming *Ghost Lab*. But if paranormal investigative work had never found a place on cable TV, we would still be doing exactly what we've done on *Ghost Lab*—traveling wherever we could to investigate all types of places alleged to be haunted, formulating new theories and then testing them out, experimenting with new technology and finding ways to apply it to paranormal phenomena, collecting and analyzing data, and tirelessly searching for answers.

4
Voices from Cell Ten

BRAD: Al "Scarface" Capone. George "Machine Gun" Kelly. Robert "Birdman" Stroud. Murderers. Sociopaths. Criminal masterminds. Alcatraz is American history's most famous prison, once home to the nation's deadliest felons. Naturally, we were fascinated by the place. We had read about some of the sinister inmates who once inhabited the island and—if the accounts of certain former guards, visitors, and paranormal groups were true—might still. A number of people swore The Rock was haunted. They heard disembodied voices and music. They felt sudden cold spots. Were malingering spirits still chained to the prison cells they had inhabited during their lives?

If you're a *Ghost Lab* fan, you might have seen our Alcatraz episode during Season One. We couldn't transport the mobile lab itself across San Francisco Bay for that investigation, but we took enough portable equipment to drop an electronic surveillance net over the former penitentiary, using recorders, video and still cameras, and data loggers. We spent two chilling nights gathering convincing evidence that the ghost stories we had been hearing were credible.

But that wasn't the first time I had set foot on The Rock. We mentioned my initial visit in the episode, but we didn't go into detail. Like my visit to Gettysburg, my night at Alcatraz had a profound effect on both my future and Barry's. In many ways, it was the impetus for the founding of Everyday Paranormal. Here's what happened.

In June 2007, I was working as a project manager for a company called Hewitt Associates, which implemented payroll and other systems for large corporations. I was assigned to the Mervyns department stores account. One of the perks of the job was that Mervyns' headquarters were in the San Francisco suburb of Hayward, California, which meant that I flew there regularly on business. On this particular trip I was working with a colleague from the Chicago office named Carol Franklin. Like me, Carol loved ghost stories and we sometimes swapped theories about the paranormal. One day we were sitting in the hotel lobby after work, trying to come up with something interesting to do. We had both visited the city so often that we had already hit many of the tourist spots.

Suddenly I got an inspiration.

"Hey," I said. "Let's go to Alcatraz!"

"We'll never get tickets," Carol replied. "I hear you have to buy them way in advance."

"You never know. Let's give it a shot."

At first the ticket agent on the phone sounded skeptical, but when she checked her computer, she said, "This must be your lucky day. There's space available on the last tour—the sunset tour. It's the most popular one."

With less than an hour to spare, we went straight to the rental car and headed for Pier 33 at Fisherman's Wharf. A few blocks before we reached it, I caught sight of a Radio Shack. On the spur of the moment, I decided to go in and buy a forty-dollar digital audio recorder.

At the time, EVPs were the newest trend in paranormal investigations. An EVP—short for Electronic Voice Phenomenon—is a voice or sound that is transmitted in a frequency too low for the human ear to detect in real time (15 hertz or below), but which a digital audio recorder is sensitive enough to catch and which becomes audible when you play the recording back. Paranormal experts, Barry and myself now included, believe that EVPs are genuine voices of disembodied spirits. Though I had never tried my hand at ghost hunting, I had been reading a lot about EVPs and I was eager to experiment with the technology. Alcatraz seemed like an ideal setting for a test run.

The late-afternoon air was crisp and laced with the scent of baking sourdough bread and the cries of gulls as we boarded the ferry with a throng of fifty or more fellow tourists. The skies were a brilliant cloudless blue, just as Gettysburg's had been. There was none of that famous San Francisco fog rolling in to lend a sinister atmosphere. But, standing on the upper deck, I got an awe-inspiring view of The Rock during our fifteen-minute voyage. There's something ominous about seeing the hulking outline of the fortress come into sharp detail as you draw near. A sense of foreboding creeps up on you. It's easy to imagine the despair prisoners must have felt at the sight of it.

The prison itself perches like a giant bird of prey on the highest point of the island. That meant we had to trudge up a long, twisting road from the wooden dock where we debarked to reach the entrance. Because the island is part of the National Park Service, green-uniformed park rangers were stationed everywhere, like a friendly modern equivalent of the onetime prison guards, to answer questions and to point us in the right direction.

"Folks, this is a special night in the history of Alcatraz," a ranger told us when we reached the front doors. "Tonight is the anniversary of the famous escape from Alcatraz. You might remember it from the movie starring Clint Eastwood."

It was June 11—thirty-five years earlier to the day when Frank Lee Morris, with brothers John and Clarence Anglin, escaped through the vents at the back of their cells, snaked along corridors hidden in the walls, shimmied down stove pipes, and threaded their way down the steep hillsides to a makeshift raft they had constructed. Then they vanished into the night. To this day, no one knows what happened to them. Some think they drowned in the treacherous currents and frigid waters. Others believe sharks ate them. But since no telltale bodies ever washed up on the island or the mainland, a number of people are convinced that they got away.

I felt a surge of excitement. If any night might be conducive to ghosts, why not this one?

Like countless other tourists, I picked up a set of headphones attached to a handheld audio cassette player and set off on my self-guided tour. It led me up and down skeletal metal staircases and in and out of five-feet-by-nine-feet steel-barred cells with chipped and peeling green paint, listening to firsthand accounts of former guards and prisoners remembering life inside The Rock.

Nobody actually got sentenced to Alcatraz, the tape told me. The Rock was where they put you if you were deemed too dangerous, too volatile, too unmanageable for other prisons to handle. There you stayed until the powers that be decided that a lower-security prison would be able to keep you under control. During Alcatraz's twenty-nine-year run as a federal penitentiary from 1934 to 1963, thirty-six prisoners tried to escape: twenty-three were caught; six were killed; two drowned; five disappeared, Morris and the Anglins among them. Eight people were murdered by inmates. Five committed suicide, and fifteen died from natural causes. The island had a morgue, but no autopsies were performed there. Nor were any bodies buried on The Rock.

I learned some lesser-known facts about the island, too—facts that might be the root of paranormal activity as easily as a former prison could. Originally built to protect San Francisco Bay after California became U.S. territory in 1848, the fort was active during the Civil War and housed Confederate prisoners. It also saw its share of misery long before that: Legend has it, thousands of years ago Native Americans banished those who broke tribal laws to the island, where they believed evil spirits dwelled. Were Alcatraz's ghosts resentful spirits of marooned tribe members doomed to life—or eternity—in isolation? One point was certain: this rocky outcropping in the Bay had seen a lot of human misery and pain over the years.

One of the theories we've developed in our research is the idea of Imprinting or Place Memory. We believe that strong memories and extreme emotions can leave a lasting mark on a place in the form of residual energy. We think this energy can influence the behavior and feelings of living visitors (making them uneasy or short of breath) and possibly set the stage for paranormal activity. Back then, we hadn't articulated a hypothesis about it; I just figured a place that had seen so much despair and rage would be a likely setting for restless spirits.

The tour led me through the warden's house, the hospital, the laundry, the library, the mess hall, and the cell blocks. Blocks B and C housed the bulk of the prison population when Alcatraz was in operation. Below these lay D Block, the old solitary confinement cells. The audio tour was intriguing, but I was beginning to lose hope of using my new recorder. There was no way I could test the device in this setting. Every time I slipped into an empty cell for a few seconds, another visitor would walk by or shuffle in next to me. Even as a novice to digital audio technology, I knew the odds of getting false positives from background noise like scraps of other people's conversation would be astronomical.

When the tour ended, I headed back to the entrance, where most of our group was already milling around. Glancing out, I realized night had fallen and it was dark outside. Even on the upper tiers, only the feeblest shards of sunlight filtered into the building. As a prisoner in the lower tiers, it must have been disorienting and depressing. The only way to tell morning from evening would have been to hear a guard bark "Wake up!" or "Lights out!"

We were getting ready to head back to the boat when one of the rangers made an announcement. "We've just learned that there's a mechanical problem with the boat. If you'll all just make your way down to the dock, you can wait there. We should be ready to depart in about forty-five minutes."

While everyone else shuffled out, grumbling about getting stuck here and missing their dinner reservations, I felt a jolt of adrenaline course through me. I waited until the crowd thinned then walked over to the park ranger stationed at Broadway—one of the intersections between cell blocks. He was sitting in his green uniform at a desk, reading a newspaper. He glanced up as I approached.

"Is there any way that I might be able to go through the main cell house one more time?" I asked. "I just want to look around the cell blocks a little more and take a few pictures while I wait for the boat." I didn't tell him I was planning to conduct an audio session or hunt for paranormal activity. I wasn't sure how he would react to a statement like that.

He shrugged. "Sure. No problem," he said. "Knock yourself out."

While Carol waited outside and took her own photos, I hurried back to the cell blocks and down the stairs to D Block. I chose "The Hole," as prisoners dubbed it long ago, because I had read about people having paranormal experiences in it. Also, I figured that if emotional torment gave rise to hauntings this would have been the epicenter of it. Inmates here were kept in their cells twenty-four hours a day in almost total darkness and given only bread and water on most days. Robert "Birdman" Stroud, a homicidal sociopath with a Jekyll-and-Hyde personality (nothing like the gentle character immortalized by actor Burt Lancaster in *The Birdman of Alcatraz*), spent six of his years at Alcatraz in solitary. Needless to say, he wasn't allowed to take any birds with him.

The place had felt desolate and oppressive enough with other tourists wandering around. It was ten times eerier knowing I was the only one here. It was cold, and a clammy dampness hung in the air. Above all, it was silent and still. I remembered reading about the rule of silence imposed throughout Alcatraz in the early 1930s, when inmates were allowed to talk to one another only at meals and recreation periods. They got so desperate to hear the sounds of other human voices that they emptied the water out of their toilets and whispered through the sewage pipes. Spending days on end in silence in The Hole must have been torture.

I stepped through the door of cell number 10. I was standing in a solid metal cube: metal roof, metal walls, metal floor. I strained my ears for sounds of the wind or birds outside, of travelers waiting at the dock, or traffic on the mainland.

Nothing.

I sat down on the floor, cleared my throat, and switched on the recorder. I knew from the shows I had seen that I needed to establish an audio base line. "This is Brad Klinge," I said awkwardly, my voice ringing in the empty room. I stated the date and explained that I was making a recording to see if anyone else might be there.

"Um . . . This is EVP session number one. I'm here in D Block, Alcatraz cell ten, solitary confinement."

Then I asked a few basic questions like the ones I had seen on paranormal TV programs.

"Is anyone here?"

"What's your name?"

"Were you a prisoner in this cell?"

"When were you here?"

It was as still and deserted as the field in Gettysburg had been on that summer day.

Today, we would use parabolic microphones and RT-EVP

recorders (supersensitive digital audio recorders that contain two microprocessors—one that records in real time or RT and one with a programmable recording delay) from the Ghost Lab to make our focus razor sharp. Sophisticated equipment like that expands your capability to analyze evidence on location as you receive it and to home in on the source of a noise with your surveillance equipment and cameras, or to catch a direct EVP response and adjust your line of questioning accordingly.

But this was old school. I was just asking questions into thin air. Talking to myself. I didn't hear anybody answering. There was no scurrying of mice or scuttling of insects. No creaking or settling of floorboards like you get in old wooden houses. I was the only living thing in the building. The closest guy was the park ranger, and he was reading in the next cell house over.

I continued the session for about thirty minutes, switching the tape on and off periodically, changing locations and asking more questions. Finally, I glanced down at my watch and realized I had better hurry to the boat or risk getting left on the island for the night.

"Man, I can't wait to get back and upload this stuff to my laptop," I told Carol enthusiastically after we climbed back into the rental car. Too anxious to wait, I switched on the recorder just to see what it sounded like as we pulled out of a parking spot.

My own voice announced, "This is Brad Klinge . . . I'm here in D Block."

Then, to my astonishment, I heard someone else say in a clear voice, "Cell Block Ten." It happened right after I said "D Block," almost as if it was correcting me.

"Did you hear that?!"

Carol nodded, wide-eyed.

I played the recording again to make sure our ears weren't tricking us.

Riveted, we listened to more of the recording. That wasn't all I had captured. A short time later came the distinct thud of footsteps, as if someone wearing heavy shoes was walking deliberately and quickly past. Then came the *slam* of a door banging shut, immediately followed by a deep resonant voice with what sounded like a Scottish accent. *"Gotcha!"* it said.

I heard myself ask, "Were you an inmate in this cell?" A thin, reedy, tremulous voice responded, "This one."

It floored me. If the Gettysburg video was visual evidence that ghosts exist, this was its audio counterpart—equally convincing aural proof. You couldn't dismiss these noises with mundane explanations like interference from radio broadcasts, passing cars, or banging pipes in a radiator. There was nobody—I mean *nobody*—in the building with me. I wracked my brain to think of ways that I might be getting duped into misinterpreting this as evidence—debunking, as paranormal investigators call it now. But I couldn't come up with any logical explanation to dismiss what was on that recording. Plus, two of these comments were specific to Alcatraz—and based on the timing and the nature of the comments, they seemed to be actual responses to my questions.

The footsteps and cell door slamming followed by the phrase, *"Gotcha!"* could have been residual hauntings—ghostly echoes of a past action that the atmosphere somehow recorded and continues to replay. But the other two were in all likelihood what we would now call intelligent hauntings—entities with the ability and the desire to interact with the living.

Was it this easy to capture EVPs?

The next night I followed my standard routine and took a walking ghost tour of the city. This one was called San Francisco Ghost Hunt. It was led by Jim Fassbinder, a shaggy-haired local ghost historian and paranormal researcher who knew a lot more about the field than I did at the time. Like other guides I had seen, he dressed in period clothing that included a top hat and long nineteenth-century-style caped black coat as he led us past "haunted" mansions, hotels, and other landmarks, regaling us with the eerie stories behind them.

After the tour I pulled Jim aside, told him about my trip to Alcatraz, and asked if he would listen to my recordings.

"Sure," he said. "I'd be happy to."

When I played them for him, he looked as astonished as I had.

"Can you think of anything else that might cause this?"

"No," he said. "And I'll tell you something more. Once in a while, I get a vision about a person. You are going to do much more in this field. You're going to be a pioneer and change the way people view the paranormal. And you're going to be known by lots and lots of people."

Did he inspire me to go into this field? Would it have happened anyway? Did he have a genuine psychic flash? Or did he simply sense my enthusiasm for the subject and the fact that I was starting to dabble in paranormal technology and "read" what was obvious about me to the careful observer? Who knows? As is sometimes the case when it comes to paranormal matters, we can only speculate.

BARRY: A few weeks after Brad got back home to Texas, we were at his house for a big family birthday party when he pulled me into his study to tell me about his trip to Alcatraz.

"You gotta hear this," he said eagerly. "Listen . . ."

I was shocked by the ghostly, echoing sounds. "Cell Block Ten" was the first EVP I had ever heard. But then the cynic in me kicked in. I started grilling Brad immediately. "Who was in there with you? Where was everybody else? Could that be somebody's conversation you overhead? Could a TV have been playing nearby and the sound carried through the walls?"

He shot down every suggestion I made. You couldn't find any easy way to dismiss this unless he had manufactured the evidence himself, and there would be no reason for Brad to go to that kind of trouble just to fool his own brother. It seemed like solid evidence to me. I was sold.

"I've been thinking of starting a paranormal team," Brad said. "Now I've got to do it!"

"Dude, I'll be your videographer!" I said.

Ironically, when we went back to Alcatraz for our *Ghost Lab* investigation, we actually used the same audio recorder that Brad bought in San Francisco back in 2007. We supplemented it with a slew of digital video cameras, remote audio recording technology, and other electronic surveillance tools, but the little forty-dollar handheld recorder was still a vital part of it all. We caught every EVP we played in that episode of *Ghost Lab* on it. In fact, we captured a lot of the evidence featured in the show's first season on that machine, and another just like it that I purchased a short time later.

We used those recorders up until Season Two. When we started Everyday Paranormal, they were basically all we had. We added some digital cameras that took still pictures, a digital thermometer, and a handheld video camera we dubbed "the Barry cam" before our first official investigation. (You'll hear more about that in the next chapter.) But we didn't have DVRs with infrared capabilities, EMF detectors, data loggers, smart board technology, elaborate computers, flat-screen TVs, or any of the other gizmos that now make up the arsenal of equipment in the Ghost Lab.

Undoubtedly, some people would argue that RT-EVPs, with their heightened sensitivity and real-time analysis capabilities, make the previous generation of audio-recording technology obsolete, but we still keep those old trusted digital audio recorders from our early days in our bags. And we use them from time to time in investigations to supplement the advanced, state-of-the-art tech when we attempt to collect EVPs.

We pride ourselves on our audio evidence. Those bell-clear Alcatraz EVPs set the bar high and we have maintained that standard ever since. We toss out any evidence we deem marginal, even if it's probably a legitimate paranormal noise, because we don't want anyone to be able to say, "Well, that *might* be a box scooting across the floor. It might be a pipe banging." Not only are we exacting when we extract data from our recordings (our standard protocol is to listen to the entire session from every investigation at least three times), but I also have a considerable background and expertise in audio. I honed my listening skills to precision during my four years in the navy, most of which I spent with headphones clamped to my head, listening to Morse code.

Why do we focus so heavily on audio? Our investigations over the years have shown that the easiest piece of paranormal evidence to gather is an EVP. We're still trying to figure out why this is the case.

The Alcatraz EVPs are what we would categorize as Class A EVPs—clearly audible and discernible on playback without digital enhancement. By contrast, Class B EVPs are relatively loud and often audible without headphones. Class C EVPs are very soft and their words may be impossible to understand. You can use software to amplify the voice and to eliminate background noise and mechanical sounds from the recorder itself, but we generally dismiss any recording that isn't at least a Class B.

Skeptics claim EVPs are generated by sources like stray radio transmissions and background noise, and that impressionable people just choose to ignore mundane explanations. Instead, they argue, the gullible imagine they hear language or other meaningful sounds in random noise. Scientists say it happens because the human brain has a natural tendency to put whatever it sees or hears into a familiar context to make sense of it. (Some scientists call it "apophenia," others "patternicity.")

But long experience has proven otherwise for us. There is no way "Cell Block Ten" was random noise, in our opinion. Not when the phrase mirrored Brad's own comment so accurately. Not when a long silence came before and after.

So, assuming EVPs do arise from paranormal sources, why do they happen in the first place? We still don't know. Is a voice the easiest way for someone in the spirit world to manifest? Does it require harnessing less energy than an apparition or a moved object? We don't know that, either. We *do* know that paranormal activity is bound by scientific laws. Using what we understand of science and continuing our research, we hope eventually to be able to answer those questions definitively.

The one thing we *did* know for certain was that we *had* to visit more places like Alcatraz and collect more EVPs like the ones from D Block. It was time to start doing our own paranormal investigative work.

To Catch a Ghost

BRAD: The Alcatraz EVPs convinced me to take my longstanding fascination with ghosts to the next level. Instead of taking paranormal tours, I decided to start conducting paranormal investigations. My initial instinct was to form my own team in San Antonio, but I ran a quick Google search beforehand to find out whether any such groups were already operating locally. Up popped the Web site for PAST (Paranormal Association of South Texas)*.

Maybe I could join them and help develop their team. I called the number on the web site and spoke with the leaders, a couple named Mary and Mark*, who agreed to let me accompany them on their next investigation. I couldn't wait. Here was a chance to get some fieldwork under my belt, to observe the methods experienced paranormal investigators use, to ask questions, and hopefully to bounce ideas off veteran ghost hunters.

Here's one notion that I was eager to run by them: After all those tours I had taken everywhere from California to eastern Canada, it didn't take a genius to note some common characteristics in the stories you hear about "real" hauntings. This suggested to me that all apparitions manifest in a similar way—that there had to be some common denominator that allowed them to communicate with the living and appear to them in certain forms under the right conditions. What was it? Could a careful, dedicated researcher collect and correlate enough data to answer the question?

I envisioned bringing new perspective and innovation to the operation—and at the outset it seemed promising. I collected some nice, clear EVPs from haunted houses in San Antonio and within two weeks I was promoted to team manager, which meant I was liaising with clients, planning out investigations, and supervising the team members who accompanied me on them. But the never-ending drama soon started to chafe. Mary and Mark bickered incessantly about who should be in charge and how their investigations ought to unfold. Worse, they were often antagonistic toward our clients. You can tell when people are genuinely freaked out about the paranormal activity they've experienced and are turning to you not out of idle curiosity but because they want help. Yet, Mark and Mary tended to ignore the anxiety of the people we were supposed to be helping and to take a condescending attitude with them. "Well, you know eighty percent of hauntings can be debunked," they would say dismissively. "You're probably imagining things." I've never bought into the notion that 80 percent of reported hauntings were bogus anyway. I've never seen any data to back up the assertion.

Mark's outsized ego irked me, too. Whenever I suggested a new idea, he shot it down instantly with a patronizing reminder that *he* had worked with the top names in the paranormal field and that was *not* how they did things.

"But that's the point!" I would say in frustration. "I want to try something new."

"Not happening," he would respond.

Keeping my mouth shut and toeing the line have never been my strong suits. Most of the time, I ignored him and experimented anyway. One night, I raised my voice during an EVP session to see if that might draw out activity.

"What are you doing?" hissed Mary in a whisper.

"Asking questions."

"You can't do it like that."

"Why not?"

"It's disrespectful."

"Why is that disrespectful?" I demanded, taken aback. It's this kind of assumption that sets my nerves on edge. What authority decreed that a boisterous or challenging approach is taboo in a paranormal investigation? Show me the data from repeated experiments indicating that a higher decibel level or a gruffer tone produces fewer EVPs or apparition photos, and I will be the first one to quiet down. But there isn't any such data. To me, the notion seemed antiquated and Victorian, borrowed from the hushed tones typical of séances and funeral parlors. Why not try a variety of voices and then analyze the recordings to determine which elicited more responses? Deep male voices? Higher female ones? Music? Laughter? Maybe we could collect enough data to draw some conclusions about what triggered activity in this particular location and then repeat the investigation to test our theory.

The PAST team kept admonishing me with the phrase, "It's always been done this way." Well, that doesn't necessarily make it the *right* way. You've got to challenge the status quo to make new discoveries. It's hard to believe closed-mindedness would be a stumbling block when it comes to exploring a topic like ghosts, but it is.

Needless to say, my suggestions went over like lead balloons. After six weeks, I decided to call it quits. I thanked PAST for the experience, but said, "I see a different way of doing things, and I'm going to try it."

I can still hear Mary's parting words: "You'll never amount to anything in this field!" It was childish, mean-spirited, and completely typical of that organization.

I held my temper and walked out the door without replying. But as far as I was concerned, she had thrown down the gauntlet. Say something like that to a competitive person like me, and I'm going to pour all my energy into proving you wrong. I was already passionate about delving deeper into this field. Her comment just galvanized my desire to make an impact on paranormal investigative work.

I drove home, went straight to the phone, and called Barry. We agreed it was time to form our own team. The only way we were going to answer the questions that kept us lying awake at night would be to research them ourselves, test our theories using whatever scientific instruments we could find, gather data, and draw our own conclusions. Sure, we would collect outside input, though from now on we would turn to real experts—seismologists, geologists, historians, psychologists, and the like.

But what should we call our group? Everyone in the field back then had an acronym because they had seen *Ghost Hunters* on Syfy and knew about TAPS (The Atlantic Paranormal Society). Paranormal groups were springing up everywhere with clever code names that mixed the notion of ghosts with words that meant something to them locally. So I started trying to think of one, too. I kicked around a bunch of ideas. How could I work in the words San Antonio? Paranormal Association of San Antonio? PASA? No. I didn't want anyone to confuse us with PAST. *Paranormal Home At The . . .* But I didn't want to call my team *PHAT-*anything.

Finally, I realized I was doing just what I condemned others for doing: I was boxing us in, limiting myself to what I had already seen. And that was precisely what I *didn't* want to do with this new group. I didn't want to follow the masses or the accepted format. So, why become another acronym-handled ghost-hunting gang following the leaders?

Our philosophy had to start with our title. If there's anything you learn in business school, it's that your company's name must tell the consumer in one word who you are. Who would our consumer be? Anyone who was having paranormal experiences, wanted to have them investigated, and came to us for help.

We would be accessible and we would focus on the *work*. No drama, no egos, no bs. I had seen too much of that already. So our name needed to reflect not only what we did, but what we were all about. We weren't geeky scientists sitting around with lab coats and Bunsen burners. We weren't Goths with black eyeliner or mystics with crystal balls and tarot cards. We had families and mortgages. We watched Monday Night Football. We were just typical American guys. Regular Joes. What about Joe Blow Paranormal? No, I could hear the bad puns already. Everyday Joe Paranormal? That didn't work, either.

How about Everyday Paranormal?

Bingo.

The word "everyday" resonated on several levels. First, paranormal activity happens every day in everyday places to everyday people when they least expect it—when they are cooking dinner or getting out of the shower or watching TV, whether it's at 3 A.M. on a Saturday or at noon on a Tuesday. Second, the word conveyed who we were and how we would approach our work—as everyday Joes. Third, I already believed that everyday science held the key to unlocking the secrets behind paranormal activity. Energy, environmental phenomena, weather—I was convinced that they were all linked to paranormal manifestations.

And though I didn't want any acronyms, if anybody ever took it upon themselves to shorten our name, Elvis Presley happens to be one of my idols so I could live with E.P. as a company nickname.

We considered operating as a nonprofit organization, but the paperwork was formidable. So instead, Barry and I opted to create a legal partnership. We filled out the forms and drove to the Guadalupe County Clerk's office at the courthouse to register the name Everyday Paranormal.

The woman working at the registration desk scanned our completed form and said, "Everyday Paranormal? What do you guys do?"

The moment we started to explain, she cut in enthusiastically. "Y'all know *this* building is haunted, don't you?" Then she launched into a story about the disembodied voices and slamming doors employees had heard.

We struck up a conversation with her and before long, a cluster of staff members had formed around us, all chiming in with their own anecdotes about encountering ghosts at the courthouse when they were working late or at the theater down the street when they were kids.

If we had worried about strangers ribbing us over the focus of our start-up business, the reaction that day quelled all our fears. The name Everyday Paranormal seemed to open people up and prompt them to share their own ghost stories even when we didn't ask.

A few weeks later we hired a friend of a friend to design T-shirts for us that we could wear while we were on investigations. We asked him to try his hand at creating a graphic treatment—nothing gory like vampires dripping blood, but nothing too cute, either. I didn't want to walk around with Casper the Friendly Ghost on my back. He came up with several options and the winner was the hooded, Grim Reaper-esque figure you still see in our logo today.

BARRY: After we got the paperwork out of the way and recruited a handful of enthusiastic friends and acquaintances to join Everyday Paranormal, we needed places to investigate. Helen Salk*, one of our female team members, knew a manager at the Harlequin Dinner Theatre at the Fort Sam Houston Army Base, so I made a call, dropped Helen's name, and explained what we wanted to do. To my delight, the manager at the Harlequin said, "Sure. Come on out next Friday."

The Harlequin had led a number of colorful past lives—first as a noncommissioned officers' (NCO) club in World War II, then as a rowdy music hall, and, since 1975, as a theater giving live performances of everything from *The Mousetrap* to *Macbeth*. If you believed the tales, some members of the old clientele had never left.

How else to explain the phantom footsteps, the laughter drifting out of empty rooms, or the shadowy bartender who vanished when patrons approached him for a drink? What about the woman who materialized to watch wordlessly from the wings when actors rehearsed, the soldier who peered out the second-floor windows, or the wizened Indian who strode purposefully across the first floor? The fabled Apache war chief Geronimo himself was held briefly on Fort Sam Houston's grounds after his capture. Could the ghost be his?

There was no history of anyone dying on the premises, but as you might have heard us explain on *Ghost Lab*, people don't have to die in a building to haunt it. We are convinced that spirits can linger in a location for the sheer love of a building or certain objects in it, because of a strong emotional connection to an event that took place there, or for many other reasons. There is also the possibility that some private tragedy took place here, which was never historically documented.

At any rate, on November 4, 2007, we told our wives and kids good night, and headed off for Everyday Paranormal's first official investigation. Darkness had fallen by the time we pulled into the parking lot, but the manager had not yet arrived to unlock the doors for us. To prep for the event, Brad and I had divided the building into three sections we wanted to focus on during the investigation. We chose them based on the locations where patrons and employees most often reported paranormal activity at the Harlequin. First was the dining and bar area; second, the theater, which included the stage, seating, and backstage dressing rooms; and third, the second floor, which comprised an office space and a long hallway with rooms on either side.

We took advantage of the downtime before being allowed inside the building to start investigating and divided everyone into teams, briefed them about where they would begin their portion of the investigation, and reviewed basic protocol. Then we milled around, leaning against our cars and making small talk to pass the time until the manager showed up.

Standing there in the dark, I wavered between surges of adrenaline-fueled anticipation and awkward, self-conscious doubt. What were we doing here? Did I really buy this stuff? What if Brad and I had misinterpreted those Alcatraz recordings and none of this was real? I glanced over at Steve Harris, who I knew was wrestling with the same skepticism. He shrugged and grinned sheepishly. Just like me, he was wondering if we were about to make fools of ourselves.

Apparently, no one else shared our uncertainty. They seemed to be chomping at the bit, overcome with curiosity and eager to get started. We had a total of about ten volunteers, more than we would include in the investigations we do nowadays. But back then we didn't want to turn away anybody who seemed enthusiastic to join us. Most of the people clustered around us outside the Harlequin were longtime friends, though a few were friends of friends who had heard about our group and asked to come along. Three in the group were self-proclaimed white witches.

We had a bare-bones arsenal back then: just a few digital cameras, a couple of digital audio recorders, and a little Sony Handycam with night-shooting capabilities that my mom had given me as a graduation present. We dubbed it "the Barry-cam." Since it belonged to me, I became the team's official videographer that night. We had also picked up a handful of inexpensive walkie-talkies and we gave one to every team member so we would be able to communicate with each other throughout the investigation and tell them when to shift from points A to B.

It wasn't long before the theater's manager, who planned to accompany us on the investigation, arrived and unlocked the doors. We decided it would be a good idea to get the lay of the land, so we asked her to show us around quickly before we set up any equipment or sent the teams off in different directions. We still do this at the beginning of every investigation. We call it the walk-through. We were on our way back to the entrance to meet the rest of the team when we suddenly caught the scent of something burning. Had one of our investigators set the place on fire in the first few minutes? We hurried toward the source of the odor and stumbled onto a bizarre sight. There were our white witches, standing in a circle, holding hands, chanting, and burning some kind of plant.

We gaped at them in unmasked surprise.

"We now raise the energy in this place," cried one in a mystical voice.

"What the hell is going on?" Brad asked.

"Come forth!" she cried in the same dramatic tone, ignoring his question.

"Excuse me," said Brad in a louder voice.

She came out of her trance and stared at us quizzically.

"What are you doing?" I asked.

"We're calling the spirits forth," she said, as if this was obvious.

"Well, we're into hard-core research here," Brad replied, working hard to hide the irritation in his voice. "That's our focus on this team, so you need to stop what you're doing *now*."

They threw us disappointed looks but dropped their hands and slouched over. We reiterated our instructions about the method the teams should follow, then sent everyone off on their missions in the pitch-blackness. I went with Brad to the main stage area, where I flipped on my camcorder and filmed Brad as he started his first EVP session, announcing his name, the date, and the location. We let the recorders run in silence for a while, listening closely to the various noises around us to help us establish an audio baseline and more accurately distinguish sounds that were normal for the old building from those that were out of the ordinary. After a few minutes, Brad started walking around, scouting the room as he tossed questions out into the darkened theater.

We had been investigating for about fifteen minutes when I slipped into a seat in one of the front rows of the theater to give my feet a rest. I was still filming Brad, who was now standing directly in front of the stage, continuing to ask questions in the hopes of prompting an EVP response from a spirit.

"Is there anybody here?"

"What is your name?"

"Did you work here?"

"Were you an actor?"

I was just beginning to wonder if all this wandering around in the dark might prove too tedious for me to enjoy every weekend when all of a sudden, someone grabbed my arm and yanked it hard. This wasn't a subtle touch. It was forceful. I whipped around to see which team member was trying silently to get my attention, but to my amazement, there was no one there. Nobody was in the room with me but Brad, and he was standing ten feet away.

"What was *that*?" I gasped.

"What?" asked Brad.

"I shit you not, but something just grabbed my arm."

My blood started pumping furiously, not out of fear but excitement. Maybe there really *was* something to this. Had I just had my first paranormal personal experience? Had something from the spirit world reached out and made physical contact with me? What other explanation could there be? We spent another ten minutes running the session in the area, tense and eager to see what would happen next. But as far as we could tell at the time, nothing did.

So we moved on to our next target: the dressing rooms. This section of the building was completely black once you switched off the flashlight beams. You had to move slowly and grope ahead of yourself or risk stumbling over furniture and bruising your shins. I switched on the camera and peered through my viewfinder, navigating the objects that were now visible only in night-vision green. Brad took a seat in one of a group of metal office chairs surrounding a big square table, set down his walkie-talkie, and switched on his recorder to begin the next session.

"Brad Klinge, EVP session two," he said into the darkness. Again, he started to ask questions.

"Is anybody in this room right now?"

"Were you a guest here?"

"Were you an officer?"

Staring through the lens, I thought I caught sight of the walkie-talkie on the table rocking back and forth subtly. But stupidly I had put the record button on pause. I swore under my breath.

"What's the matter?" Brad asked.

"I think I might've just missed something," I muttered.

Resolving not to let it happen again, I focused my video screen on the walkie-talkie and waited without moving a muscle. About thirty seconds ticked by, and then the empty chair next to Brad suddenly flew away from the table as if an invisible hand had shoved it.

Brad, who couldn't see what had happened but heard a loud metallic scraping sound right next to him, jumped about a foot in the air and leaped to his feet.

"I caught it! I caught it on video!" I screamed.

We hit Replay on the camcorder and watched the footage. Sure enough, there it was—a heavy, four-legged office chair without wheels moving nearly three feet across the floor all by itself.

Later, we did everything we could to try and re-create the chair's movement. Could someone have kicked it? Could the floor have been uneven? Could one of us have bumped something else that hit the chair and caused it to move? Try as we might, we couldn't stage a scenario that would make the action recur.

This is still standard operating procedure for us: No matter how convincing evidence seems, we always take it with a grain of salt. We try to find other possible explanations and rule them out one by one, as should any self-respecting paranormal investigator. If at last we conclude that there is no possible mundane, nonparanormal explanation for what happened, we admit it as genuine evidence.

As the night wore on, a number of our team members had what we now call personal experiences. They heard noises. They felt odd sensations. They thought they glimpsed figures flitting through the darkness. But it wasn't until we reviewed the evidence next day that we realized just how much data we had actually collected.

As I mentioned, I continued to film after being grabbed during Brad's first EVP session in the theater. Roughly five minutes after my experience, the footage on the Barry-cam showed a clearly defined shadow person—a black figure about the size of an adult man—who appeared to stand up in front of me and walk off to the right out of the camera's frame and around the stage. It appeared as if he had been sitting there with me, watching Brad, but then abruptly decided to get up and leave. And I'm not talking about fleeting footage that you might miss if you blinked; it was a long, clear sequence.

This struck us as significant—two pieces of evidence that seemed to correlate. I was glad I had had the presence of mind to keep filming after I got grabbed. As we have learned over the years, activity tends to happen in bursts. You have to shoot first and ask questions later. If you get a personal experience, you can't panic or start jabbering about it to everyone else in the room. You need to grab whatever technology you have on hand and get it in gear quickly. Take readings on your EMF detector—a handheld scientific instrument that measures electromagnetic fields and registers sudden surges in the power level, which paranormal investigators associate with the presence of a ghost. That way, you'll be able to compare the numbers to your baseline readings to find out whether EMF is spiking. (Unfortunately, we didn't have one yet when we investigated the Harlequin.) Turn on your video camera and your audio recorder. Ask questions aloud. Focus all your attention on getting more evidence to correlate with what just happened to you. Otherwise, it's nothing more than a personal experience. It might make a good ghost story for a cold night around the campfire, but no investigator worth his salt is going to deem it genuine proof of paranormal activity.

In reviewing the evidence, we also discovered that the photographs we had collected were as striking as the video footage. One of the apparitions frequently described by patrons and employees of the Harlequin was of a young girl in a frilly, old-fashioned outfit. Sure enough, Steve Harris caught a clearly defined image of a child in a pale blue, puffy-sleeved dress. Another photo revealed a half-naked woman reflected in a mirror. We double-checked to confirm that this wasn't simply a reflection of a painting or a bust in the theater, but the building contained no similar artwork. Finally, we got a great picture of the famous phantom bartender. We were standing in the bar area in the dark, snapping flash photographs of what seemed to be an empty room. In one frame, a man appears, wearing a tuxedo and bow tie with what looks like a towel hanging over his arm.

These weren't shadows or mist. They were clearly defined full-color pictures. People still look at them and say, "Who's the guy in this picture?" and "Where'd that little girl come from?"

Paranormal investigators still aren't quite sure why digital cameras and video recorders pick up apparitions that the human eye can't see. We do know that the technology provides a much greater depth of field than traditional film cameras do and that it reacts differently to light than the human eye does. One theory on apparition photos is that ghosts tend to manifest only in the infrared part of the color spectrum, which normal vision can't discern. Another is that apparitions move too quickly through our environment for us to see.

Of course, not all evidence that people claim to catch on camera is legit. One of the classic misinterpretations concerns "orbs." Dozens of haunted places we visited in our ghost tour days featured walls lined with photographs visitors had sent in, claiming the hazy circles appearing to float in midair were proof of paranormal activity. Sorry to disappoint everyone, but that's just dust and moisture. We've done experiments to prove this. If you take a rug and shake it and then snap a picture moments later, you'll get the same effect—dozens of floating orbs. *How come it's in this picture and not in that one?* people demand. Because dust floats. When an air conditioner turns on, it sends lots of dust particles into the air. Shutting or opening a window can create the same effect. So can walking across a dusty floor. Our motto is, if it looks and behaves exactly like dust, then you have to throw it out as evidence because you will never know if it's dust or if it's a ghost.

After reviewing the evidence we collected at the Harlequin, we did what fans have seen us do in every episode of *Ghost Lab*. We sat down with our client—in this case, the Harlequin's manager and owner—and reviewed the evidence with them. They were blown away.

"You're not the first paranormal group to come through here," the owner of the Harlequin told us. "But no one has collected evidence like this."

That incident still resonates. We had provided them with real, pertinent data including color photographs and a video sequence—not just a tape of banging shutters that might have been explained away by a strong wind. We call it "holy shit evidence." We couldn't pinpoint what was causing it, but we concluded from the wealth of data that the Harlequin was indeed experiencing genuine paranormal activity. For a business long reputed to be haunted and a staff curious to know whether there was any truth to the tales, we had provided a service.

We walked away feeling energized and eager for more. To date, the Harlequin still ranks as one of our most productive investigations and the one that set the standard for Everyday Paranormal's visual evidence.

Something stuck in our craw, though. We were adamant about staying true to our name and mission. And there was nothing down-to-earth or everyday about white witches burning sage. They would have to go.

Before the next investigation, we cut our ties with them. Two of them parted ways graciously, but the third vowed to put a curse on us. "Isn't that *black* magic rather than white?" we asked before hanging up on her. It wouldn't be the last time someone threatened to hex me, or jinx me, or put a curse on me. You don't have to work in this field long to realize that some of your weirdest encounters are bound to be with the living—not the dead.

We continued to allow psychics to tag along on our investigations for the next few months, though we warned them to leave the mumbo jumbo at home. "When you investigate with us, you are just a regular person," we told them. "We don't want to hear that you're getting a sense that there's a child standing in the corner or a voice from the beyond is commanding you to go upstairs. Keep that to yourself. Show us the photos, the videos, and the voices on recorders."

From that moment forward, I took over as case manager. I would cold-call different places, introduce myself, and explain how we operated, then ask if we could investigate their premises. "Hey, we heard you have a ghost story," I would say. "Can we come investigate?" Nine times out ten, the response would be a curt "No thanks" or a surprised hesitation, followed by, "Uh... Yeah... Let me get right back to you on that, okay?" Then I would never hear from them again.

But that tenth time, I would get a "Sure, come on out!" And that's what mattered. That's all it took to find some incredible evidence.

Chilling Evidence from an Ice Rink

BRAD AND BARRY: The Freeman Coliseum is one of San Antonio's unofficial landmarks. Built in the late 1940s, the arena and the grounds that ring it have hosted countless rodeos, fairs, circuses, rock concerts, trade shows, and other events. Elvis Presley played here in 1956. Roy Rogers and Dale Evans hosted the first nationally televised rodeo here the following year. The Harlem Globetrotters, KISS, and countless other headliners of their day performed at the Freeman, too.

Over the years, the place witnessed a lot of human drama and high emotion. Victories. Defeats. Wishes fulfilled. Dreams destroyed. It also set the stage for some tragedy. Ten deaths occurred in the Coliseum over the years, including a clown dropping dead of a heart attack in the midst of a performance.

You can trace the site's dramatic history back even further. Long before the arena was built, the area was part of the Fort Sam Houston Army Base extension. In 1889, it set the stage for the San Antonio International Fair and Exposition. Less than thirty years later, the barns allegedly held World War I German POWs, who made shoes in a makeshift workshop. Graffiti scrawled in German remained scratched onto the walls for decades afterward. Teddy

Roosevelt trained his fabled Rough Riders, the first U.S. Volunteer Cavalry, on these grounds before they rode off to battle in the Spanish-American War. The Buffalo Soldiers who accompanied them set up an encampment on the property, too.

When we were kids, going to the Freeman was a huge deal. Whenever you went there, you could count on seeing something exciting and memorable—like the Alzafar Shrine Circus, Sesame Street on Ice, and, of course, the San Antonio Stock Show and Rodeo. We attended the rodeo every February as a family. We would watch bull riding, barrel racing, and cattle roping in the main arena, then head outdoors into the sun to wander through the fairgrounds, enjoying the carnival rides and cotton candy. As schoolchildren, we piled onto yellow busses for class trips to the Freeman, where we toured the barns to learn about the animals and watch the cattle auctions. The Stock Show and Rodeo was famous; more than a million visitors flocked to it every year. In fact, the place was essentially built to host that particular event back in the days when San Antonio was first and foremost a cattle town. When we got older, we still headed to the Freeman—to watch monster-truck shows and rock concerts.

Then in 2003, the SBC Center (now the AT&T Center) sprang up next door, offering fancier digs and overshadowing the Freeman, which found itself relegated to hosting only intermittent second-tier shows. All at once, the venerable place looked shabby, outdated, and forlorn.

Little did we know our most exciting and memorable experiences at the Freeman were still to come—during the night Everyday Paranormal spent investigating the old arena to find out whether the many rumors about hauntings there were true.

BRAD: As I mentioned, I learned to drive a Zamboni as a college kid when I got a weekend job at San Antonio's Crystal Ice Palace, a local ice rink unrelated to the Freeman. Bill Oberle, the guy who taught me to handle the massive ice-smoothing machine without crashing it into the boards or hacking up the ice, became a close friend of mine. When I graduated from college in 1996, I said good-bye to Bill, turned in my keys, and figured I would never resurface ice again.

I had gotten married in 1993 and, shortly after earning my degree, I moved to Dallas with my wife, Jessica. But a year and a half later, we were ready to head back home. We returned to San Antonio in 1998, shortly after our first daughter was born, happy to be closer to family and friends. It wasn't long before I started running into the guys from the Crystal Ice Palace around town. Now they all seemed to be working with the Central Hockey League's San Antonio Iguanas, who practiced and played their home games at the Freeman. One day, my old mentor Bill said, "Hey, I could use some help at the rink. You interested?"

I already had a full-time job, but I agreed to start splitting the Iguanas' games, youth hockey league tournaments, and other skating events with Bill on the side. There I was, a twenty-four-year-old college grad, spending Friday nights and Saturday mornings back on the ice. At least I was helping an old friend out.

I was blindsided a short time later when Bill was rushed to the hospital with heart problems. He was barely forty years old when he died a few weeks later of a congenital heart problem that none of his other colleagues, myself included, knew about.

Not only was Bill's death a blow, but my moonlighting workload doubled overnight. I suddenly found myself scrambling to do all the ice resurfacing and putting in as many as thirty hours a week in hockey season, between October and April, on top of my regular job. I was definitely sleep-deprived, but that didn't explain the bizarre events that started happening in the old arena.

The Freeman is on East Houston Street, hugging the outskirts of the city's dicey eastern edge—a neighborhood where you wouldn't want to be stranded late at night. Being there alone gave you a creepy, uneasy feeling. The place was cavernous, hollow, and felt acutely empty when you were there by yourself. I kept my mind focused on my work: heave the rattling metal garage door up, get the water in the Zamboni . . . and resist the urge to glance apprehensively over my shoulder.

One night after a game ended and the arena emptied out, I was alone in the building going over the ice to get it ready for hockey practice the next morning. Then something high up in the stands caught my eye. Only the work lights were on, which meant it was dimmer than it would be for a game but not dark. I glanced up and spied a man standing on a seat, watching me silently, his hands thrust deep in his pockets.

A series of rapid-fire thoughts flashed through my mind.

The first was a panicked, *Who the hell is that? Did that guy break in or sneak through a door that some careless person left open?*

Immediately following that came a flood of relief at recognizing the man's blue windbreaker and khaki pants—the unofficial "uniform" Bill wore to work every day. I hadn't seen the face clearly, but I would know that cheap blue windbreaker anywhere.

Reassured, I turned my attention back to the wheel to navigate a corner, and in that fraction of a second, the hair on the back of my neck stood up. Bill was dead. He had died almost a year earlier.

I whipped my head around again, staring wildly back up into the stands.

The figure was gone.

The entire incident unfolded in a matter of seconds. There was no way anybody could have sprinted out of the stands, no way anyone could have ducked down to crouch behind a seat and hide from me.

I had just seen a ghost—most likely the ghost of my late friend Bill.

Less than a month later, I was working around 10:30 in the morning in the deserted arena, getting it ready to open at noon for a tournament. The only other person on the grounds was the security guard who had let my car through the west gate, and he was where I had left him, out at the edge of the parking lot manning the booth. As I walked down the hallway that wraps around the arena, past closed concession stands and locked restrooms, I heard the sound of running footsteps approaching—feet pounding hard over the concrete floor. Once again, my initial thought was that someone who wasn't supposed to be here had sneaked in. I ducked into an alcove and peered around the corner. To my surprise, the footsteps grew louder and louder, as if someone were approaching and passing me, then faded away as if the person were running on down the hall. Yet nobody appeared. "What the hell?" I muttered. I walked into the main arena. It was deserted. Something invisible had just given me a perfect demonstration of the Doppler effect . . . but what?

Not long after this, I was working in the machine shop when I heard a metallic noise across the room. I glanced toward the source of the sound: a nearby table on which a collection of nuts and bolts was lying. As I watched, several of them rose into the air and fell back onto the table with loud clunks, just as if an invisible hand had scooped them up and let them fall.

Something strange was going on at the Freeman. Was the place haunted? When I finally broached the topic with some of the guys I knew who had worked there for years, to my surprise they answered almost nonchalantly. "Sure," several of them said with a shrug. "Weird shit happens here pretty often." These were tough, no-nonsense guys who worked security, maintenance, and grounds. But almost every one of them had had personal experiences with the paranormal at the Freeman. They heard disembodied footsteps and voices. They saw doors slam and seats flip down by themselves. They caught fleeting glimpses of movement—as if figures were darting past—in the arena and the barns when nobody was around. A number of them had heard screams. Several had seen a woman who looked like the late head of the Stock Show sitting in a box seat, smoking a cigarette. They had approached her and found no one—only a strong smell of cigarettes hanging in the air. Steve Harris started working at the Freeman around the time I did, and the same kinds of unnerving, inexplicable things happened to him. Eventually, the employees stopped trying to find logical explanations and chalked up what they witnessed as an eccentricity of the arena.

I started researching the history of the Freeman and talked with the general manager, who had worked there for decades. He filled me in on its colorful past and little-known secrets—like the fact that a network of tunnels (by that point, disused and dangerous due to crumbling walls and water seepage) spider-webbed under the entire property. And the fact that in 1918, the Fair and Exposition Building that stood here burned to the ground. He also showed me the German graffiti in the barns.

BARRY: When we first started looking for sites to investigate, the Freeman seemed like a natural. We knew its giant proportions would present some challenges, but that would broaden our field experience. Brad and Steve Harris were now driving the Zam for the American Hockey League's San Antonio Rampage, the farm team for the National Hockey League's Phoenix Coyotes. Unfortunately, they played at the AT&T Center, which meant we no longer had an "in" at the Freeman. Worse, the arena was owned by Bexar County, which meant we might face months of frustrating red tape or run up against a county commissioner who wouldn't like the notion of a bunch of ghost chasers running around the facility after hours in the dark. We were ready to rule out the whole idea when a guy named Andrew, who had recently become a member of the Everyday Paranormal team, mentioned that he knew someone on the Freeman grounds crew. He made one call and we were in.

The investigation was set for 10 P.M. on a Friday night in December 2007, late enough so that the parking lot the Freeman shared with the AT&T Center would have emptied out after the hockey game.

We figured it might be fun to film a video of ourselves investigating and upload it to YouTube, so we posted an ad on craigslist for a videographer. Our requirements were simple: basic camera skills, reasonable knowledge of film-editing software, and a willingness to join a one-night paranormal investigation without expecting to be paid. We got a slew of responses from local companies, the kind who film weddings and corporate events, promising us a slick product for a small fortune. We also got an e-mail from a local company called Ace Productions* that would later play a memorable role in the evolution of Everyday Paranormal. But that's another story. We'll fill you in on it later.

The winner was a guy named Mike Berger. He became a cameraman on our first season of *Ghost Lab,* and he is now an Emmy-winning editor, but back then he was a kid fresh out of his teens and film school in Orlando. Brad met him briefly at a restaurant a few days before the Freeman investigation; the rest of us didn't set eyes on him until we pulled our cars into the parking lot that night.

Brad and I arrived to find the Freeman looking decidedly shabby and worse for wear. Aside from the occasional B-list rock band, nobody played here anymore, and the arena spent most of its time deserted. As our half dozen team members stepped out of their cars, the first comment on everyone's lips was, "Whoa, this place is huge. How are we going to cover it?"

We explained the plan we had mapped out beforehand, dividing everyone into teams as we had done at the Harlequin, and instructing them to each target a different hot spot (an area reputed to have paranormal activity)—the outer ring where Brad heard the footsteps, the offices, and the main floor with its surrounding seating. We had acquired a couple of digital thermometers, but aside from that, we were still working with the basic equipment we used at the Harlequin: digital cameras, our trusty audio recorders, and my camcorder, which I now handed over to Mike. It was the first time we had been on camera for anything other than a home movie. We tried to ignore our "filmmaker" as he walked around us, tracking our movements. I had gotten used to being the man behind the camera on our last investigation, so it felt weird knowing somebody was zooming in on me for a close-up. We cracked a few jokes about starring in a TV show, but we never dreamed that our offhand comments would foreshadow real events.

Mike trailed us as Andrew's buddy unlocked the doors and we stepped into the cavernous, vacant interior of the arena. The emergency lights lent a dim greenish glow to the halls and the main floor, making it hard to see anything at first. As we would soon discover, only certain sections of the Freeman were illuminated. The dressing rooms and offices were pitch-black.

I stood there waiting for my eyes to adjust to the darkness and wondering if the place would live up to all the stories I had been hearing. Was it really haunted? Within the first hour, I started to suspect the answer to that question was a resounding *yes*.

We split into teams and headed off in our separate directions, planning to touch base in about twenty minutes. I was working with Brad and two other guys on the main floor. The first strategy we tried was to remain quiet and motionless in one spot, and wait to see whether any activity occurred. We all stretched out in a row on the floor in the center of the auditorium, resisting the urge to talk and instead just soaking in the ambiance. But then a voice came crackling across the walkie-talkie, breaking the silence.

"This is Team B. We're at the top of the seating area right now. How many of you are lying there?"

"Four of us," I replied. "Why?"

"We're counting five."

"Negative. We've got four down here."

"Well, everybody up here is counting five."

That was just for starters.

All night long, team members had personal experiences. They felt sudden cold spots where no air-conditioning vents were located. They heard footsteps when no one was moving. The most intriguing experience involved a guy named Claudio, who worked for the Rampage hockey team and knew Andrew's buddy on the grounds crew. Curious to see a ghost hunt up close, he had asked to come along with us. He was heading down one of the hallways with another of our novice investigators when a man crossed their path a few feet ahead and walked out the bank of glass doors. They followed him, only to discover to their astonishment that the doors were locked and there was nobody outside them.

"What did he look like?" Brad demanded when Claudio told him.

"I didn't see his face, but he was wearing a blue windbreaker and tan pants," came the answer. Neither of these people had ever heard Brad's stories about Bill Oberle.

A few hours later, we were sitting in the dressing-room area on the lower level, about twenty feet from the spot where legend says a panicked bull once broke free from his pen and trampled a young girl to death during a rodeo. Mike Berger was filming Brad as he sat Indian style with his walkie-talkie on his lap. All of a sudden, the radio went flying off his leg. We couldn't see it in the darkness—Brad just felt the device move and I heard a clatter—

but the camcorder had an I/R (infrared) illuminator, so Mike saw it clearly as it happened.

At another point in the night several of us distinctly heard a woman scream. We radioed the other two groups to make sure it wasn't one of them. Rather than wait to replay the recording in the lab, we decided to check it out on the spot to confirm that we had captured the scream. As soon as we heard the sound play back clearly, we attempted to pinpoint the location based on its volume. We surmised that it had risen out of one of the stairwells, so we moved there to investigate further. It turned out to be the same section where the girl had supposedly been crushed. That event helped us to recognize the value of real-time analysis and to make it a regular part of our investigations.

Of course, we can't do all our analysis in real time. In some ways, the investigation doesn't truly get underway until we reach the lab and start combing through all the hours of recordings. I won't lie. It's arduous and it's tedious. But it pays off. We didn't have a formal lab in the early days; we just uploaded everything to our own home computers. And in this case the recordings yielded a veritable gold mine of evidence, including an apparition image that remains one of our best to date and which correlated to a sighting we would otherwise have written off as no more than an interesting personal experience.

We were standing in the area where Brad had heard the running footsteps, near a snack bar and storage area. Nothing seemed to be happening except for Brad's camera shutter clicking as he snapped pictures of the hallway, aiming the lens in various directions. Suddenly, he did a double take and whispered, "Did you see that? Somebody just walked by!"

I didn't catch it that time, but I did before the night ended. We were in the process of wrapping things up and had gathered in the center of the arena to recap. We were just about to walk out when I happened to glance up at the seating sections to our right. Each door was punctuated by a square of faint light where an exit was located. I distinctly saw a person walk from left to right through one of those blocks of light. It looked like a black figure, but that could have been explained by backlighting, which tends to make everything look cloaked in black. I turned around and counted everyone to find out who was still wandering around in the halls alone. When I realized that every member of our team was accounted for, I thought, *I might have just seen my first apparition.*

We examined Brad's photos from the area near the snack bar—the place where he had heard the running footsteps years earlier—and we discovered a remarkable sequence that seemed to correspond to the figure we had both seen: The first picture was perfectly clear and detailed. The second was blurry, and in the center of the blur was the shape of an enormous man. He looked to be about seven feet tall. Where his feet should have been, his body became wispy and tapered into the floor like a genie emerging from a bottle. The third picture, taken immediately after this, was sharp and clear, just like the first. And Brad had caught this series of photos at just about the same time he had glimpsed something moving through the darkness.

As for the EVPs, they were loud and crystal-clear. Listening to the recording from the session in the dressing room, we heard someone say, "Hi," in a long, drawn-out, feminine voice. It was half seductive, half downright chilling. And it happened in the same time frame that the radio flew off Brad's lap.

We also caught one of the creepiest EVPs any of us had ever heard. It came from our investigation of Section 25, near the top of the arena, a spot where employees often encountered odd events. Brad was kidding around, saying things like, "Ooooooo, all these stories about the Freeman Coliseum being haunted. Yeah, right. I'm so scared. You know what? I'm not buying it. I think it's bullshit." It was probably the first time we had ever tried what's known in the field as "provoking."

As if in answer to Brad's snide commentary, the recording revealed a long, deep-throated, cackling laughter—a sort of demonic "HAHAHAHAHAHA!!!" that harked back to maniacal clowns from the horror movies we watched as kids. It remains the scariest laugh I have ever heard. When we play that EVP for people, they all get the chills.

Our provoking techniques emerged spontaneously out of our cynical, sarcastic senses of humor—in other words, out of us just being ourselves. And they worked. So the proverbial lightbulb went on. We said, "Aha! Something we did elicited a response. Maybe it was the tone of voice, maybe the volume, maybe the nature of the comments. Whatever the trigger, it might be effective. Let's continue to test it and collect more data." We did and, though it isn't failsafe, we *have* collected a lot of EVPs using provoking since then. We're still not exactly sure why it works well for us. It might be rooted in the fact that aggression, anger, and combativeness generate more energy (in the form of adrenaline and stress hormones like cortisol) than calmer human emotions do. Maybe ghosts tap into that energy source and channel it into a vocal response.

The third and fourth EVPs we caught revealed a female

voice saying, "Can't you help me?" on the arena's main floor and a male voice in the hallway saying, "Ask if they're real." The latter really intrigued us because it played into the theory of parallel dimensions—a concept that had fascinated us for a long time and that we thought might dovetail with ghost activity. This was the first EVP we had ever collected that seemed to be referring to us. Did it indicate that ghosts were seeing us? Were they confused by our presence? Did they think *we* were the ghosts? (If you've ever seen the 2001 horror movie *The Others,* you'll know what I'm talking about. If you haven't seen it, rent it. It's one of our favorites.)

Our night at the Freeman not only helped us to formulate techniques that we still use regularly today, but it gave us solid material to mine for our theories about how the paranormal operates. We started to formulate hypotheses about the possibility of parallel dimensions and how, when, and why they might overlap or collide. Over time, as we continued to collect evidence, we were able to categorize paranormal evidence based on what we had seen in past investigations. Amassing data sometimes reinforced our theories; other times it led us in new directions.

We've heard some pretty weird EVPs in the past few years. Some of the words seem so random or ridiculous that we have both shaken our heads and said, "Man, ghosts say the stupidest things sometimes. Don't they?" But that in itself gives rise to interesting questions: Would a ghost intentionally spout gibberish? If so, why? Do the words have meaning to the dead that the living can't decipher, like a code we haven't yet cracked? Or could we be catching scraps of conversations that would make perfect sense if we could only collect the missing pieces and put them all in context?

On the other hand, some of the evidence we have analyzed seems anything but random. Take the ghost in the blue windbreaker, for example. We thought a lot about him. Here's one theory we discussed: If you were a spirit attempting to communicate with the living, and you knew you could harness enough energy to manifest for just a few seconds, what might you do? You might try to give clues to your identity. You might say a name—your own or that of someone important in your life. Could this explain the prevalence of EVPs that include names? Or you might give another visual or verbal clue to your identity. Bill's blue jacket was like his trademark. It identified him instantly to his friends at the arena. To this day, Brad is convinced that the apparition he glimpsed from the Zamboni was Bill's—and that his old friend was trying to get the message through that he was still around, at least in spirit.

7
Echoes in an Abandoned Old Folks' Home

BARRY: The cops in Leon Valley were getting fed up. Calls had been coming in to 911 from a nursing home on the outskirts of this bedroom community of San Antonio for months. No one ever actually reported an emergency. In fact, the operators never even heard a voice on the other end of the line—just hang-ups or silence—but the police were still compelled to respond to every call to "clear" it.

The bizarre part of all this was the fact that the care center had closed a year earlier. The power had been turned off, the doors locked, and the phone lines disconnected months ago. That meant the 911 calls couldn't be written off as pranks, wrong numbers, or even malfunctioning wires. The telephone company had looked into the matter and confirmed that the number was indeed out of service and had never been reassigned. They were at as much of a loss as everyone else to explain who could be making the calls or how they kept getting through.

For the officers on duty, driving out to the forlorn one-story building was bad enough. Having to venture inside was worse. The place gave them all the heebie-jeebies. No one would set foot inside alone and, even with a partner, walking down the shadowy halls and peering into the darkened rooms was unnerving. There was something grim, sinister, and almost surreal about the abandoned facility. Half-empty glasses of moldy orange juice still stood on bedside tables. Wheelchairs sat empty—not lined neatly against the walls, but facing in odd, random directions along the halls. File drawers had been pulled open and medical records were strewn over the floor all around the front desk. The chaos left behind suggested that the entire population, staff and patients, had rushed out in a panic one night and never bothered to go back to get their personal effects or clean up. And in fact, that was essentially what had happened: The state shut the center down for various code violations, and the people in charge closed up shop quickly and then vanished. No one seemed to know what had become of the residents, though they must have been moved to other care centers in the area.

Vandals might have broken in at some point and added to the jumble, dumping the records on the floor as they searched for credit card numbers and other personal information. Kids could have snuck in and raced the wheelchairs up and down the halls, crashing them and leaving them lying on their sides. But none of that could explain the bizarre experiences the cops had whenever they entered the deserted nursing home. Objects moved when nobody was touching them. Noises echoed from the far end of the halls. Officers often felt the disturbing sensation of someone standing behind them and peering over their shoulder. They would glance back quickly but find no one there. These were all veteran cops who didn't get spooked easily, and they were all convinced that the building was haunted.

Brad's brother-in-law, Anthony, was one of the Leon Valley

police officers who responded to calls at the care center. One afternoon, as he and another cop canvassed the nursing home, checking yet again for intruders after a 911 call, every door lining the corridor in which they stood slammed shut simultaneously. There were about sixteen doors altogether, and the tremendous *bang* made the walls shudder. These were old-fashioned doors—not automatic doors that might have been triggered by a short in old wiring. The two cops stared at each other in shock and then tore out of the building.

"It's one of the spookiest places I've ever set foot in," Anthony told Brad. "It sure seems like it's haunted. You ought to check it out."

When it comes to paranormal investigations, my motto has always been the creepier the better. Anthony got permission for us to explore the building from the chief of police, and we made plans to investigate as soon as possible—late one Friday night just before the Christmas holidays in 2007.

Shortly before he left home that evening, Brad got a call from Felicity*, one of the psychics who accompanied us on our investigations sporadically. She knew about the investigation, but wasn't taking part in it.

"I'm worried about you," she told him urgently. "I'm sensing that they know you're coming out there. You need to put your protective bubble on because you're in danger."

"My *what*?" Brad asked.

"Your protective bubble."

"Okay," he said, puzzled. "Exactly who do you think knows I'm coming?"

"I'm looking at them . . . I can picture them . . . and they're . . . *really old.*"

Really? Well, that's a no-brainer, Brad thought. *We're going to an old folks' home.* He thanked her and hung up. Was it genuine concern? A legitimate premonition? We were beginning to suspect that Felicity and the other psychics we had encountered thrived on heightening the drama whenever possible. We loaded our equipment into the cars and thought little more about the warning.

It was around midnight and bitterly cold as Brad and I steered down the narrow weed-choked road that led to the vacant care center. The building was well off the beaten path, on the fringes of a residential neighborhood, the houses dark and quiet at this late hour. The parking lot was empty. Behind it, the grounds were unkempt and overgrown as if no one bothered to maintain them anymore. Even looking at it from outside, you got an ominous, foreboding feeling about the building. I guessed that most people would avoid coming here even in broad daylight.

In addition to our core team of Steve, Hector, Jason, and videographer Mike Berger, we had invited a few friends of friends who were eager to take part in one of our investigations, and while we waited for them to arrive, we stomped our feet and blew on our hands to warm them. It was one of the coldest December nights I could remember. At last, they showed up. Anthony opened the doors for us, and we got our first glimpse of the building's interior. He hadn't been exaggerating. The scene in front of us looked like it had come straight out of a movie set. Wheelchairs littered the halls, some facing the wall or blocking doorways, others overturned.

Disused medical equipment was everywhere. Records, charts, and handwritten pages were scattered across the carpet as if someone had turned the folders upside down and shaken their contents out deliberately.

The temperature inside was hardly warmer than it was outside. You could see your breath. Shining our flashlights around to get the lay of the land, we realized that the building was designed in the shape of a cross, with four wings radiating from a central nurses' station where a large desk stood. Peering down the hallways, we could see open doorways leading into darkened bedrooms on either side. It was easy to understand why hearing all those doors slam unexpectedly at once would send a person flying for the nearest exit.

We set up our monitoring equipment in the hub and the four halls. We had our standard walkie-talkies so we would be able to communicate with each other as well as cell phones. (They were turned off so they wouldn't interfere with our equipment or generate false positives, but we'd have them on hand in case of emergency.) Incidentally, we had realized through our first few forays in ghost hunting that high-tech devices aren't the only useful equipment to bring along. Extra batteries for your flashlights and recorders are indispensable. So are notepads, pens or pencils, a tape measure, and a luminous watch. I still firmly believe that the best piece of equipment you have is yourself—your own eyes and ears—but you need gadgets as reinforcement so that you will be able to record key details of any unusual occurrence as soon as it happens. It's smart to have a first-aid kit on hand in the car, too. You never know what you might bump into or trip over when you're exploring a strange place in the dark.

BRAD: As usual, we divided into teams—four of them in this case so that we could each search a different hallway and the rooms that adjoined it. My group was conducting an EVP session in one of the empty bedrooms when we got our first inkling that the strange encounters the cops had described were not the result of overactive imagination. Three of us were asking questions aloud, talking to each other, and generally making our presence known to see whether we could draw out any activity. Suddenly, Barry felt a tightness in his legs, as if someone or something was wringing his pant legs as hard as they could. He didn't say anything at the time, but he took a few steps across the room to see if he could shake off the sensation. Instead, it intensified and a vague nausea began to creep over him. Was he imagining it or did the cold air in this room seem heavier and more oppressive than it had in the hallway?

Standing next to him was a woman named Kristin*—one of the people who had come along to experience her first ghost hunt. "Man," she said abruptly. "My legs feel really tight."

"No shit?" Barry asked. "Because that's exactly what's happening to *my* legs right now."

"Um . . . can we get out of here?" she asked.

As if in response to her question, the bathroom door suddenly slammed shut with an enormous *crash*. The noise scared the hell out of all three of us. No matter how many ghost investigations you've been on, a loud noise two feet away when you're standing in pitch-blackness is enough to trigger anyone's flight response. We resisted the urge to bolt out of the room and instead turned our attention to the door and the small bathroom beyond it, shining our flashlights into every corner and continuing to record our work. We hadn't felt a gust of wind, but all the same we checked to make sure there were no broken or open windows near us. We could find nothing that would have caused the door to slam. It looked like object manipulation to me. (Object manipulation, in ghost investigators' parlance, involves any inanimate object being moved by an unseen force. Whenever an object appears to move of its own accord, you have to rule out gravity, magnetic forces, air currents, and other invisible but perfectly logical causes first, of course. If and when you can do that, you've got some pretty compelling and unusual paranormal evidence.) We were convinced there was a paranormal entity in that bedroom, and its presence was so intense that we were relieved to get back out into the hall and away from it. Once we did, the queasy, stifling feeling dissipated immediately.

We completed our rounds of the bedrooms in our wing and then reconvened with the larger group in the hub to compare notes. We were standing in a circle near the nurses' station talking when something small and hard skipped across the floor between our legs, narrowly missing us before it shattered against the wall.

"What was *that*?!" several people blurted out at once.

We walked over to get a closer look at the missile. Lying on the floor were the remains of a small lightbulb. A bit of probing with flashlight beams revealed that the ceiling was lined with light fixtures, each containing the same type of bulb and each held in place by a metal cover. Could a bulb have fallen out? We scanned the fixtures and the floor to see whether any of the covers might have come loose, but none had. Every one of them had a bulb and a cover in place. Besides, this bulb hadn't fallen. It had come zinging across the ground, parallel to it, and slammed up against the wall as if an unseen hand had whipped it at us.

No one could have removed a cover, unscrewed a lightbulb, and put it back without drawing our attention. Had an extra bulb been lying in the clutter on the floor? Maybe. Could someone have chucked it as a prank? Sure . . . except for the fact that our entire group had been huddled together. As always, we made sure everyone was present and accounted for so that we could document the incident accurately. Then we calculated the trajectory. No one had been standing near the spot from where the bulb had been hurled.

Most of the newcomers in our group were already pretty well spooked, but as if to convince any remaining skeptics that there were inhuman presences nearby, strange inexplicable noises began to echo down the halls. Doors slammed in deserted parts of the building. An eerie *sshhhhh-shhhhh* hissed from the empty residents' rooms, as if the metal-ringed dividing curtains still hanging between the metal beds were being pushed back and forth. Members of the team felt cold spots and the same unsettling conviction the cops had experienced of someone standing behind them. Anthony was with us, in uniform and on duty throughout the investigation, and at one point I was filming him when he flinched and turned to peer intently down the dark hallway.

"What is it?" I asked.

"I don't know," he said. "I thought I heard someone down there."

When we replayed the footage the next day, it gave us all a

shock: a split second before he spun around, there was the sound of an old woman screaming. It was the kind of evidence we love to find—a personal experience that we can correlate with recorded paranormal data. How many people have thought at some point in their lives that they caught a glimpse of something flitting around a corner or heard a whisper when they were alone in the house? When and if it happens, most of us write it off. *We're tired. We're on edge. We must have imagined it,* we say. But the data we collected that night at the old folks' home suggested to us that, at least in some instances, those feelings are *not* pure imagination. They might be legitimate responses to paranormal activity that we sense on a visceral level.

What we captured bordered between EVP and VP (voice phenomenon)—an unexplained voice that's actually audible to the human ear in real time. It intrigued us that though Anthony claimed the noise was so faint that he barely heard it, on the videotape it was loud. We already knew about variations in frequency levels—and that the recorder captured paranormal sounds too low in frequency for the human ear to detect. But were they all at the same low frequency? Was there a specific frequency level associated with paranormal activity? Did the frequency depend on setting and/or conditions like temperature and humidity? Might it vary depending on the voice the spirit had when it was alive? For example, would the frequency a child's ghost used to communicate vary from that which an adult male's ghost used? If we could figure out the right frequency level for a given setting, would we be able to communicate effectively with ghosts?

We haven't isolated a single frequency used by ghosts, but we have conducted experiments to compare the frequencies of EVPs to those of human voices. For example, in our first episode in Season One of *Ghost Lab*, shortly after we caught a Class A EVP at the Shreveport Municipal Auditorium in Louisiana, we asked team member Steve Harris to stand in the spot where we had captured the EVP and repeat the phrase we heard on the EVP ("They saw the light."), matching his voice as closely as he could to the recorded one. Then we studied representations of both voices on computer to analyze their pitch (sound) and amplitude (signal power).

To give you some basic background on sound: Sounds are made up of evenly spaced waves of air molecules. The shorter the wavelength—or distance from one wave peak to the next—the higher the sound. The longer the wavelength, the lower the sound. Sounds with longer wavelengths don't reach your ear as frequently—hence, a lower frequency. Frequencies between 30 and 300 kilohertz (kHz) are classified as "low frequency." Perfect human hearing is 20 hertz to 20 kilohertz and average human hearing is well above the low range, which explains why EVPs aren't always heard: They occur in the lower frequencies.

We looked at an analysis graph of each voice—Steve's and the EVP—and realized that the frequency of the human voice and that of the mystery voice were entirely different when each one uttered the phrase, "They saw the light." The unidentified voice we captured represented a much lower frequency—too low even for a man with a deep bass voice. Data like this helps us to distinguish genuine paranormal voice phenomena from extraneous human voices we might accidentally catch, demonstrates to dubious viewers that our evidence has not been manufactured, and gives us data to compare with other EVPs we've collected in other locations.

The scream wasn't the only chilling EVP we captured on video camera that night in Leon Valley. The most unsettling occurred as we walked out of the bedroom where the slamming bathroom door caught us off guard: On camera you see us exit the bedroom and, just as we step back into the hallway, a male voice distinctly whispers, "Behind you!"

Over the years, Barry and I have ventured into some spots that would make the hair on the back of your neck stand up, from mortuaries to mental wards. But the old folks' home at the end of that deserted road was as creepy as any of them. Years of experience have convinced us that ghosts are complex, just like living humans. You can't easily categorize them all as either malevolent or benevolent, just as you can't always categorize people as "bad" or "good." But whatever was lingering in that desolate former nursing home was, if not downright menacing, at least tortured. Maybe the building had seen so much human suffering and misery that the hauntings, whether residual or intelligent, retained that anguish. There are some scary statistics out there about nursing homes—between 30 and 40 percent of patients suffer from malnutrition, dehydration, or abuse every year; 66 percent suffer from dementia; and 20 percent are physically restrained at some point. When you combine that data with the fact that the state shuttered this particular center as a result of its business practices, it's a good bet that when it was operational there were some unhappy souls trapped in that place, before or after death. Then again, there's a possibility that the hauntings predated the nursing home and that the staff and residents experienced the same kinds of bizarre activity that we did when we visited.

Paranormal investigators theorize that powerful emotions, negative or positive, can give rise to hauntings. But could there be some other impetus for paranormal activity in a nursing home? Are ghosts more likely to appear in a place where people die? Is that why we hear about so many hauntings in hospitals, morgues, and battlefields? Are they transition points between this world and the next, where spirits might get confused or stuck? Or, do all the ghost stories set in places like graveyards stem from superstition? Maybe we all have a predilection for believing places closely associated with death are haunted because those places scare us. Are people inherently afraid of them because they are afraid of death? We don't have definitive answers to those questions yet, but we are collecting data on the types of places where paranormal activity occurs most frequently. Once we've accumulated enough information to develop solid hypotheses, we'll let you know.

Incidentally, though people tend to associate cemeteries with ghosts, we've found them pretty disappointing when it comes to evidence collection. We've investigated quite a few graveyards over the years, but—with the exception of one shadow person in Tombstone, Arizona—we've come up empty-handed. In terms of paranormal activity, most are a dead zone—*literally*.

If we were to investigate the same nursing home now, we would do some research on the surrounding area to ensure that there weren't any mundane explanations we might be overlooking. For instance, unsealed landfills leaching chemicals can create noxious odors that cause nausea. High radon levels can trigger unusual physical sensations—say, a sense of heaviness or tightness in your legs.

As you know, our theory is that ghosts need energy to manifest and that this is why you tend to see and hear them in places with high electromagnetic fields. Today, if we were going to go back to the nursing home, we would check the EMF levels in various parts of the building. If they were abnormally high, as they often are in haunted places, we would try to pinpoint the source of the EMF. Is there a power plant operating nearby? Does a polluted river run behind the building? Turns out, dirty water is a huge conductor of electricity. Is something else acting as a battery, powering up EVPs and object manipulation? We have found that spirits can lie dormant like a volcano and then suddenly erupt when we investigate. We think it happens because empty, dilapidated places are devoid of energy—until we bring in a bunch of cameras and other technology and leave them running all night. Essentially, we've brought in the battery that allows the activity to occur. Our own energy and emotions might add fuel to the fire. Is that what we did on that bone-chilling night back in 2007? Did we create a ghost buffet?

Studies show that certain individuals are hypersensitive to EMF, which means they can end up feeling sick when they enter a high-EMF field. Were we standing in a field like that when the bathroom door crashed shut behind us? If so, it might explain the stomach-turning sensation Barry and Kristin got just before the door slammed. But, of course, back in 2007, we were just starting out. We had yet to raise the funds for EMF meters and the other sophisticated devices we use nowadays.

We were so intrigued by the haunted old folks' home that we

would have been tempted to return to film an episode of *Ghost Lab* there, but about a year after our investigation the building was finally demolished. Nobody in Leon Valley was particularly sad to see it go, from what we heard. Nor were they surprised when the land stayed vacant. Maybe even prospective developers were wary to build there again and risk disturbing whatever restless spirits still roamed the grounds.

Mysteries at Myrtles

BARRY: The rambling antebellum Myrtles Plantation in St. Francisville, Louisiana, dates from 1796 and is often cited as America's most haunted house. Now an eleven-room bed-and-breakfast listed on the National Register of Historic Places, the Myrtles has been featured by everyone from *The New York Times* to *National Geographic Explorer* for its mysterious ambiance and sinister legends.

The original house was built by General David Bradford, who was sentenced to death for his part in the Whiskey Rebellion, a violent tax protest staged by Pennsylvania farmers in the 1790s. George Washington himself signed Bradford's execution order, but somehow "Whiskey Dave" managed to evade the authorities, head south, and live out his days in style. He wasn't the only Myrtles's owner with a few skeletons in his closet. According to legend, his successor Clark Woodruff caught a slave named Chloe, who may or may not have been his mistress, eavesdropping outside a door and cut off her ear as punishment. She wore a green turban to hide it every day after that until, as the story goes, she was hanged and her body thrown into the Mississippi River. Some sources claim she managed to poison a few members of the household with oleander first. Chloe's ghost is said to wander the grounds even now, still wearing her green headdress. And she's got a lot of company from the spirit world, if you believe the staff and visitors.

There are literally dozens of ghost stories associated with the Myrtles. When it comes to paranormal activity, you name it, they've got it—from disembodied voices to full-bodied apparitions. Guests hear children's laughter ringing from the upstairs bedrooms. They feel playful spirits bouncing on the four-poster canopied beds. Some people have been so distraught by the sounds of a baby crying that they have followed the noise to a closed bedroom door and knocked to see if they might be able to help—only to find out the next day that the room was unoccupied when they heard the sobs. Visitors to the Ladies' Parlor on the first floor often feel children tugging on their clothing as if trying to get their attention, and female guests discover that their earrings were removed while they were walking through the room. The popular belief is that the ghosts belong to the children of Ruffin Gray Stirling and his wife, Mary Catherine Cobb, who owned the mansion in the mid-nineteenth century and raised nine children there, five of whom died young.

Ominous stories are told about murders of servants, children, and soldiers, though admittedly it's a little hard to sort out fact from fiction. The Union army did ransack the plantation during the Civil War and the house served as a makeshift headquarters for officers when battles were raging nearby. But the only documented murder victim is William Winter, who owned the plantation until an unknown assailant rode up on horseback one night in 1871, called him out to the porch, and shot him for reasons that remain unclear to this day. Winter staggered up the stairs and collapsed, dying in his wife's arms. Lots of people claim to hear heavy faltering footfalls on the staircase at night—noises they think are generated by Winter's ghost retracing his last steps. To add to the intrigue, the Myrtles is allegedly built on an Indian burial ground and some visitors claim to have glimpsed an apparition of a young Native American woman.

We had known about the Myrtles since 2002, when *Unsolved Mysteries* devoted an episode to the place, and it topped our wish list of targets for investigation. As soon as Everyday Paranormal was up and running, we started planning a visit. Until then our investigations had all been local, so this would be a big trip for us. To make the most of it, we decided to promote the investigation in San Antonio and let anyone who might be eager for a hands-on tutorial in ghost trekking sign up to accompany us on a first-come, first-served basis. We charged a small fee, which covered the cost of our guests' rooms for one night and asked them to provide their own transportation. The approach served a dual purpose. First, we wanted to help bring ghost hunting into the mainstream, and inviting John Q. Public along was a great way to encourage and educate others who shared our interest in the paranormal. Second, the Myrtles is so big and sprawling that we needed all the help we could find to cover it. We got a healthy enough response to our offer to book the entire B&B for a night in February 2008. Counting our crew and our guests, our group would total about twenty people.

We headed out of San Antonio early Friday afternoon and rolled into St. Francisville eight hours later, where we caught our first glimpse of the famous Myrtles. It certainly looked like a haunted mansion—part grandeur, part gloom. A wide shady veranda ran the length of the facade. A spacious courtyard stretched out in back. Inside, the decor harked back to the Old South, with its long velvet curtains, crystal chandeliers, and hand-painted stained-glass windows. Our guests wouldn't be disappointed in their surroundings. The place earned an A for atmosphere. We could only hope the hauntings would prove as authentic as the antiques.

Because it is crucial to go into every investigation armed with as much knowledge as possible, we always sit down with the people who live or work in the setting and interview them. We ask them to tell us everything they know about the history of the place, to describe anything unusual they have experienced, and to share any anecdotes they have heard from other people about ghost encounters. Both the general manager and the caretaker were kind enough to spend time talking with us about paranormal activity in the house. They shared a wealth of information about their own encounters with ghosts and numerous tales they had heard from guests.

After our conversation, Brad and I conducted a preliminary walk-through of the house and grounds. Along the way we set up equipment, including our DVRs with infrared (IR) imaging and our newly added thermal-imaging cameras, which would run all night in the guest bedrooms—a point we notified everyone about in advance. (Paranormal theory holds that some ghosts are unable to materialize in the range of the color spectrum the human eye can see, but that they are "visible" in the low-frequency, long-wave infrared portion of the spectrum—in other words, the section of the spectrum that we detect only as heat or radiated energy. Though the two technologies are slightly different, both IR and thermal-imaging technologies capture apparitions in essentially the same way: They take color-coded pictures and videos of the environment's temperature that show the variations in it. The spectrum ranges from black for the coldest temps to white for the hottest. So, if a ghost's temperature varies from its surroundings, it should show up in recorded thermal-imaging data.)

Because there's a ghost story attached to nearly every corner of the Myrtles, we wanted to monitor it all. To simplify the investigation, we broke the property into six main areas: the caretaker's cabin on the north end of the grounds; the museum on the south end of the main house (including the parlors and the staircase); the four upstairs bedrooms on the north end of the house; the Bradford Suite on the ground floor adjacent to the back porch; the carriage house or slave quarters on the south end, including its own four guest bedrooms; and finally the pond and gazebo on the west side of the property. Our groups would rotate from station to station so that everyone got a chance to monitor all of the key areas at some point during the night.

Adrenaline pumping, we assembled everyone on the first floor, divided into teams, and started our investigations at 9 P.M. Things got off to a slow start for my group. We sat in the darkness straining our ears and hoping to hear the sounds of a boot on the stairs, but aside from our own breathing and intermittent movements as one person or another shifted to a more comfortable position, there was silence. I radioed Brad, Hector, and the other team leaders to check in. Their units, too, were coming up empty-handed so far. Had we dragged everybody here for nothing? What if the whole night proved unproductive? It would be a long ride back and a big disappointment to our neophyte ghost hunters.

We plowed on, the hours dragging by with nothing happening. The occasional hoot of an owl or a pop and groan of an old floorboard settling would prompt us to sit up swiftly and peer attentively into the darkness for a few seconds before realizing dejectedly that we were just hearing normal nighttime noises.

"You never see this stuff on TV, do you?" I quipped to lighten the mood and break the monotony. "You only see the exciting stuff—not the hours of sitting in the dark. But that's the reality of a ghost investigation."

Then, as the hands of the clock crept toward 2 A.M., a dramatic change occurred. It seemed as if a wake-up call audible only to the spirit world had sounded, and paranormal activity started bursting out everywhere.

We were standing on the steps where William Winter took his last gasps when we felt the temperature around us plummet. A massive cold spot moved steadily down the staircase and enveloped each of us in turn. Fascinated, we checked the luminous screens on our digital thermometers. We were all getting 47 to 48 degrees. Moments earlier, the ambient temperature had been near 80 in the house. That meant the staircase was experiencing a dramatic and instantaneous mobile cold spell—almost a 40-degree plunge. We marked the time: 2:15 A.M. Moments later, the frigid air dissipated as quickly as it had come. It wasn't long until I became aware of an unusual odor that seemed to have wafted down the stairway along with the chill.

"Can you smell that?" I asked.

"Cigars, right?" said one of our guests.

"I smell it, too," another agreed. "And there's something else. Something like flowers."

"Are either of you wearing perfume?" I asked the women in our group. They shook their heads.

"Are you carrying it in your purses? Maybe a bottle spilled."

"No," they assured me.

"Let's see if it's confined to the staircase," I suggested. We walked down to the first floor and peered out at the porch to make sure nobody from another one of our investigative teams had ducked out for a smoking break. Nobody had. As we stood in the foyer, we became gradually aware of the cloyingly sweet scent again. We scouted around for bouquets, room fresheners, or potpourri but found none. Nor were there any flowering shrubs outside the windows that might have explained the strong fragrance. Puzzled, we moved into the adjoining parlor. So did the perfume. Each time we moved into a new room, we thought we had left the fragrance behind, only to catch a whiff of flowers a few seconds later as if the scented cloud was trailing us from room to room. The longer we stood in one spot, the more intense it grew. It wasn't until we walked outside that it evaporated.

As we exited through the foyer, which reminded me of a museum with its immaculate original furnishings and antique rugs, I passed a piano near the door. On a whim, I leaned over and hit one of the keys three times. *Ding, ding, ding.* All of a sudden something skittered across the wide planks of the wooden floor—*ch-ch-ch-ch-ch*—and slammed into the banister of the stairs.

"What was *that*?!" I exclaimed. The others, standing on the veranda waiting for me, peered in quizzically. I bent down and directed my flashlight beam at the area next to the stairs. To my surprise, I spied a rock about the size of a quarter lying there. Just like in Leon Valley, I made calculations and determined the trajectory. It had come from the Ladies' Parlor. What the heck would a rock be doing inside the Myrtles? Who threw it when the place was empty?

Was this a hostile gesture from the paranormal world? Had I breached the decorum of a bygone era and angered a ghost by tapping on the piano? Was the flying rock a random paranormal response to a sudden loud noise or maybe an attempt to communicate? Brad and I gave it a lot of thought afterward. Could you project the typical motivations and emotions of a living person onto a ghost's actions? People usually throw rocks in anger or self-defense. They do it to attack or to protect themselves. Once in a while, we throw pebbles to get someone's attention. But should we assume that ghosts' motivations mirror ours?

In a perfect world, we would set up an experiment and try different approaches—like an aggressive verbal tone or a calm, polite one—to see which resulted in a hurled rock most often. But paranormal activity is way too unpredictable for that. We've only had objects thrown at us a handful of times in hundreds of investigations over the years. That's not enough to allow us to gather adequate data. From what we can tell, when objects do come flying at us out of the darkness, it happens irrespective of the investigative technique we were using at the time. At the old folks' home, the lightbulb narrowly missed us when we were just chatting. This time, it was a response to a random, impulsive action on my part. I had no particular intent or emotion in mind when I hit the keys. Does this suggest that the action has more to do with the spirit in question or possibly the energy level in the environment than it does with the investigator's technique? We're not sure. We're still trying to find out whether certain types of ghosts are more likely to manipulate objects than others or whether any ghost will do so if it can harness enough energy.

BRAD: Speaking of antagonizing ghosts enough to elicit a response, at the same time Barry nearly got hit by the rock, my team and I were doing our best to provoke the spirits in the old slave quarters. They seemed to be ignoring our taunts and verbal challenges, so we finally gave up and decided to move on to another part of the property. As I stepped out the door, I felt someone strike me hard in the back between my shoulder blades with what felt like an open palm. I'm a big guy, but the force of it almost knocked me off my feet. I lost my balance and went careening into the three guys in front of me.

Thinking someone was playing a trick, I whirled around and realized there was no one behind me. "Man, I just got hit *hard*," I said. I lifted up my shirt and sure enough the others saw a wide red mark as if I had been smacked. Later, Barry videotaped it for documentation before it had a chance to fade. My skin was still stinging as I stalked away, furious, indignant, and unnerved. From the moment we had entered the slave quarters, I had been feeling uneasy—as if we were unwelcome here and were being watched. But as you know, I go ballistic when other people tell me about the "vibes" they're getting from an investigation site, so I kept my hunch to myself. It was only a personal experience—not the hard data we demand in the Ghost Lab—but that blow left little doubt in my mind that something in the building resented our presence enough to literally push us out.

After that, I was apprehensive about what the team members bunking in the former slave quarters that night would encounter, but the worst that happened was mild. All four returned to their rooms to find them blazing hot, the thermostats cranked up as high as they would go, despite the staff's assurances that the heat had been turned off. Sound mundane? Consider this: The building was famous for the raging fire that nearly demolished it a century earlier.

Barry and I had chosen to share a guest room in the main house. We weren't planning to sleep, just to pause for a short, uncomfortable rest in our work clothes and shoes after we wrapped up the investigation around 4 A.M. But neither of us shut our eyes. We were too wound up and wary after having each been singled out for what we thought might be antagonistic actions. We weren't quite sure these ghosts wanted the Klinge brothers in their house.

And we weren't the only ones who couldn't sleep. You could look out the window and see people lugging their suitcases across the courtyard to knock on doors where other guests they knew were staying. It got to be almost comical. Nobody wanted to stay in a room alone.

When the sun finally rose the next morning we all gathered for a country breakfast in the dining room during which we shared ghost stories and compared notes. One of our novice ghost trekkers swore that something had climbed onto the bed in which he and his wife were sleeping. Its weight made the mattress sag, he said. He felt it crawl over the bed and then wriggle itself in between the two of them. His wife, who had slept soundly through the whole encounter, turned pale when she heard the story. Another foursome had lodged in a small building known as the caretaker's cottage. We called it Poppy's cabin, in honor of a grumpy former caretaker whose ghost was sometimes seen by passersby brandishing his fist and warning them to clear off, as cantankerous after his death as before it. They were all horsing around, blowing off steam between 4 and 5 A.M. One guy knocked another's baseball cap off his head. Exhausted and preoccupied with his wrestling match, the owner of the cap never bothered to pick it up. So it was with considerable surprise that he had awakened that morning to find it hanging from his bedpost. The other three men in the cabin swore they had not touched the cap.

We were all feeling drained after the exhilaration of the night before, but we couldn't wait to analyze our data. As soon as we hit the road for the drive home, the passengers in all the cars whipped out laptops and started watching their videos or listening to their recordings. Had any of us managed to collect hard evidence to back up the myriad personal experiences?

Sure enough, we had. One team caught a Class A EVP on the porch, in which a man shouts angrily, "Get out here!" We thought instantly of William Winter. Could it be the voice of his killer? You'll remember from my discussion of Alcatraz that Barry and I believe traumatic events can leave an imprint on their environment and create a residual haunting—a ghostly playback of the original action, sort of like a paranormal echo. Some repeat themselves nightly, some yearly on the anniversary of the original action, some irregularly. We thought that this EVP and the famous staggering footfalls so many guests have heard on the staircase at the Myrtles suggested that Winter's murder might have created a residual haunting.

A session in a bedroom on the first floor of the house, which we dubbed the Chloe Suite, produced another great EVP. In it, the voice of a female or a young boy asks, "Colonel, aren't you afraid?" At first, we assumed the question referred to a Civil War soldier—perhaps to the apparition of a tall man in a gray uniform and hat that people sometimes glimpse around the gazebo. However, the Myrtles staff told us that a young girl who lived in the house in the late 1800s was infamous for her bossiness and fondness for tormenting her little brothers and sisters. Their nickname for her? The colonel.

And that wasn't all. Apparitions on video are among the rarest forms of paranormal data. Most of us will be lucky to collect a handful of them during a lifetime of ghost hunting. To our astonishment, three showed up from our hours at the Myrtles—all on the same camera.

The Myrtles caretaker told us that she sometimes walked into the upstairs bedroom where a little girl had died of yellow fever years before to find the dolls rearranged. We thought it would be amazing if we could catch one moving, so we set up a DVR with night-shooting capabilities facing the doll shelf near the door and left it running. Hidden in the hours of footage was one of the most amazing phenomena we had ever seen. Someone accidentally left a light on in the hallway and it cast a large patch of gold across the doorway. At around 8 P.M., the tape revealed a black shadow moving into the illuminated area of the doorway and standing there, almost as if it was peeking around the edge of the door to see who was inside the room. We checked our notes and our tapes and confirmed that no one had been upstairs in the house when this happened. Most amazing, the recording revealed the shadow person apparently shutting off the hall light as it exited. Our first thought was that the light was on a timer. But we inspected the fixture later and realized that it was controlled by a manual switch. As if this weren't enough, at about 3 A.M., a similar shadow about the size of a human child moved into the room, appeared to walk around, and then slid across the bed where one of the guests was sleeping. (You should have seen her face when she watched the video the next day!)

We had another camera running in the room where the shadow appeared, and when we synched the footage from the two cameras and watched them on a split screen, we noticed an interesting anomaly. At the exact moment when the shadow appeared on camera one, camera two suddenly experienced some sort of electrical interference that caused the picture to dissolve into gray-and-black horizontal lines momentarily. When the shadow vanished in camera one, the picture returned to normal in camera two. As you know, we believe entities can use electricity to manifest. Did this data reinforce that theory? Could the electrical surge from the manifestation have caused the other camera to malfunction?

We thanked everyone who took part in the project and congratulated them on their findings. Then we called the Myrtles' staff to go over the evidence with them and to thank them again for giving us the opportunity to investigate. Everyday Paranormal was far from the first team to hunt for ghosts at America's most haunted residence, but we are proud to say that Barry's rock now resides in a glass case in the Ladies' Parlor next to the many earrings female guests have lost to the resident ghosts.

We were so mesmerized by the haunted mansion that we returned in May 2008, with our team and another group of novice investigators to tackle the mysteries of the Myrtles a second time. Most of the first-timers signed up after hearing us speak at the Texas Folklife Festival, which is organized by the Institute of Texan Cultures and takes place every June. (We'll tell you more about the Institute and the investigation we conducted there later.) Our presentation was a runaway success that surprised us even more than it surprised the Institute. It also caught the eye of a guy named D. B.*, who ran a local production company he called Ace Productions.* While we were making the final arrangements for the second trip to Myrtles, D. B. called us and asked if we would be interested in working with him and his business partner on a DVD that he said might lead to a TV development deal. We remembered his name because he had responded to our craigslist ad for a videographer before the Freeman investigation. We were intrigued enough to meet with them and then to invite them along as our guests for the Myrtles investigation.

We also brought along Steve Hock, the head administrator at the school where Barry taught. Steve had recently learned about our sideline and wanted to get a taste of ghost hunting himself. What better place to start than the Myrtles? He liked it so much that he became a full-time member of the Everyday Paranormal team (he retired from his job with the school district at the end of Season One), and his deep mellow voice turned out to be perfect for voice-overs. He provided all the historical narrative for the DVDs we eventually made of our early investigations, and his old-time Southern storyteller style gave them just the right amount of drama and suspense.

To our delight, the second trip proved nearly as fruitful as the first. Again, we interviewed the staff to find out about any new phenomena from the past few months. Then we visited the local graveyard where several of the plantation's former owners are buried.

As always we ran EVP sessions and asked questions, targeting them to what we knew about the ghosts of the Myrtles.

"Who's here? Is it William Winter? Are you a slave who used to take care of this place? A child? A Confederate soldier? A Union soldier?"

"What year do you think it is? It's 2008, not 1865."

"Are you alone? Or are you with people?"

Again, we experienced dramatic spikes and drops in temperature on the stairs. The same inexplicable odors of cigar smoke and perfume trailed us inside the main house. I was perched on a settee in the Gentlemen's Parlor, conducting an EVP session, when I heard its legs creak and distinctly felt the weight of someone sitting down next to me. It gave me cold chills. We also heard loud disembodied footsteps on the porch. Even better, our first-timers caught an EVP with what sounded like a black-powder gun firing on the porch. This evidence, too, dovetailed with our theory about a residual haunting inspired by the Winter murder.

We also encountered an intriguing voice phenomenon (VP). It happened as I was strolling around the gazebo with two investigators from our Louisiana affiliate. Suddenly all three of us heard a loud male voice talking animatedly. We stopped, straining our ears to make out the words, but the pond nearby is full of frogs, and their nonstop chorus made it impossible to understand the man's words. We canvassed the area but found nobody there. Unfortunately, we didn't pick up anything but bullfrog croaks on the recordings, so we had no evidence to support the VP.

This time, Barry and I had elected to stay in Poppy's cabin with a few of the other guys on our team. We dropped our bags there and returned to the main house briefly to make sure our guests were situated. When we got back to our rooms to start setting up our equipment, we found all the bedcovers ripped off and rumpled into heaps. None of the other guests had keys, and the staff vowed they hadn't set foot in the cabin. Unfortunately, it happened before the cameras started rolling, so we failed to capture video evidence to back up what we saw. Intriguing as these last few incidents were, they were personal experiences—not hard data that would pass muster today.

We returned to the Myrtles a third time to film an episode for Season One of *Ghost Lab*. That night we tested a new technique we had developed and dubbed Era Cues—audio and visual stimuli that re-create the time period when the energy in a place was highest in order to draw out a response from spirits who lived during that particular era. We invited a local historian to read aloud from soldiers' letters written during the Civil War. It helped us to unearth even more intriguing data, including a VP of the children's voices that Barry and I had both heard during our initial walk-through of the William Winter's Suite (recordings picked the noise up, too) and yet another cold spot on the stairs that coincided with an EVP that said, "Why can't you help me?"

The old Louisiana plantation remains to this day one of our favorite places and we recommend a visit for anyone fascinated by the paranormal. We can't guarantee a ghost sighting, of course, but your odds are as good there as anyplace else you're likely to find.

Ghosts of the Garden District

BRAD: As we were preparing for our first trip to the Myrtles, we got a call from a woman named Maxine*, who lived in New Orleans's Garden District. She had seen our Web site and wanted us to help her figure out who or what was haunting the building she owned. Specifically, she wanted us to investigate two condominiums—hers and the one she rented to a local party planner named Kevin*. Both Maxine and Kevin had heard people talking when there was nobody around and they had seen a man in a Confederate officer's uniform standing in the hall at one time or another. Even more unusual, Kevin kept discovering antique coins on the floor of his bathroom.

Since it wasn't too far out of the way, we decided to swing by the Big Easy on our way home from the Myrtles. Maxine's building turned out to be a small but elegant townhouse on a picturesque street lined with live oaks, wrought-iron fences, and the opulent nineteenth-century mansions typical of the Garden District.

It was also right across the street from the famed Lafayette Cemetery Number One, with its rows and rows of crumbling above-ground crypts. New Orleans's oldest planned cemetery, it dates officially from 1833, though records show that there were burials here as early as 1824, when the ground was part of the Livaudais family's plantation in the City of Lafayette. After New Orleans annexed Lafayette and established a formal graveyard, hundreds of victims of yellow fever epidemics were laid to rest here, some whose bodies were simply left at the gates by relatives who couldn't afford a burial, due partly to the fact that coffin dealers were price-gouging like crazy. (In the summer of 1853, the city's death rate soared to a formidable one in fifteen. More than 12,000 people succumbed to yellow fever in New Orleans that year and Louisiana earned the dubious distinction of having the highest annual death rate of any state during the entire nineteenth century.) As time went by, the graveyard fell into ruins and suffered heavy vandalism for decades before a dedicated save-our-cemeteries effort restored it.

Incidentally, raised tombs became the norm in New Orleans because of the city's shallow groundwater due to its proximity to sea level. It didn't take much rain to fill graves with water and cause the coffins in them to float to the surface. A heavy storm could literally raise the dead. An influx of French, Spanish, and Italian settlers in the eighteenth and nineteenth centuries also helped to popularize raised crypts that created "cities of the dead." The style had been fashionable in the Mediterranean for thousands of years, largely because the region's rocky soil made aboveground burial so much easier than digging graves.

Maxine accompanied us on our walk-through, starting with the small foyer that led up a winding staircase to several upper floors. Along the way, she told us a little of the building's history. A bordello had apparently stood on the site in the nineteenth century until it caught fire and burned to the ground, killing several prostitutes.

After touring the property and gathering some more background information from the pair of them, to our surprise, Maxine and Kevin handed over their keys. "We'll see you tomorrow morning," they said. "Can't wait to hear what you find out!" We were flattered. Not only did complete strangers trust Everyday Paranormal with their homes and their personal belongings, but they took our investigation seriously enough to clear out of our way so that we would be able to work without distractions. They didn't hang around goggling or asking questions, as if our main purpose was to entertain them. We appreciated that attitude, and we resolved to do our best to unearth the answers they wanted.

Speaking of unearthing, during the walk-through Maxine mentioned a point that intrigued us: she was having some repairs made to the house's foundation and workers had been using a crawl space below the building. In the course of their work, they had come across a section of crumbled stone that they thought might be part of an old and long-forgotten crypt. Dark, airless, and claustrophia-inducing, the crawl space gave the men the unnerving sensation of being buried alive every time they squeezed into it. As if that weren't bad enough, bizarre things happened to them when they were wedged into the earthen compartment. They would sense something moving around them in the blackness and suddenly become so gripped with terror that they felt paralyzed. They had been in other tight subterranean spaces before, but none had left them feeling so spooked. There was something almost evil about that crawl space, they said. A few of them refused to work on the condo because they were so scared of it.

"Would you say the paranormal activity in your house has intensified since the work started?" we asked Maxine.

"You know, I think it has," she answered, surprised.

We thought what Maxine, Kevin, and the workers were experiencing might be an example of what we call Renovation Theory: the idea that overhauling a building can alter its energy level and disrupt spirits inside it, acting as a catalyst for sudden bursts of paranormal activity. (Some paranormal researchers even believe the bumps in the night are literally confused ghosts stumbling over areas where walls have been moved, doors have been added, and other sudden changes have been made in the environment after many years without disruption.)

We knew we needed to get somebody down into that crawl space with a video camera to collect data. Neither Barry nor I would ever send members of our team anyplace we would be afraid to go, so naturally, we were the first pair to volunteer. You can't be a ghost hunter if you can't deal with claustrophia, heights, getting locked into a cell, or sitting alone in the dark. It's not a job for the faint of heart. But I'm six feet four inches, 300 pounds. Barry is six feet two, 250 pounds. This came down to simple mathematics. Try as we might, neither of us could squeeze into the narrow three-by-one-foot opening to the crawl space.

I scanned the team standing around us, staring down at the black hole under the house. Who would fit into it? Mike was the skinniest. "Dude, whaddya think?" I said to him. It was like drawing the short straw. He agreed, but it was clear this would not be his favorite assignment since joining us the previous winter.

We shoved the poor guy into the hole headfirst on his belly with his video camera. He wriggled himself painstakingly down under the house until all that was sticking out were his legs and shoes. A minute or two elapsed and then all of a sudden, he started freaking out, shouting and scrambling frantically to back himself out.

"There's something down there! I saw it!" he spluttered as he emerged, caked in dirt and sweat. "Some kind of shadow thing started crawling toward me, and then it disappeared. It looked just like that thing in *The Grudge,* all disjointed and freaky!"

Back out in the daylight, he gradually calmed down, though he made it clear there was no way he would ever venture into that crawl space again. We had worked with Mike for months and we knew he didn't scare easily. Still, we wondered about the power of suggestion. Everyone had been talking about how haunted the crawl space was. Maybe the guy was claustrophobic. Maybe he had recently seen a horror movie and images from it were still fresh in his mind, waiting for the right cue to trigger them. What if those factors combined when he found himself under the house? Panic can kick anyone's imagination into overdrive. That's why you need hard data to back up personal experiences.

But then we reviewed the evidence and we started to think Mike's reaction might have been more rational than we suspected initially. On the videotape a split second before he started to yell and struggle so desperately to get out of the crawl space, there was a distinct sound of somebody whistling sharply and then a rock came flying across the screen of the video.

It was intriguing evidence, but we needed to move on to the condo and get both apartments and the hallway wired for sound and video. Dusk was falling and we were about halfway through the job when we ran short on cable. "There's more in the truck," I told the rest of the crew. "I'll run down and get it." I dashed downstairs and was crossing the foyer when I remembered I had given Barry the keys to the truck. I did a 180, intending to race back upstairs to grab the keys from him. To my surprise, I saw a woman standing in front of me about an arm's length away, wearing a long red dress and an elaborate matching hat. I almost ran right into her.

"Excuse me, ma'am," I said, stepping to the side to avoid bowling her over. I wondered what on earth she could be doing here in the foyer. Maybe she was an actress dressed in old-fashioned clothes for a performance in one of the events Kevin was planning. If so, she probably didn't realize that Kevin was out for the night. I turned back to ask her, but she was gone.

"What the . . ." I blurted out. No matter how often you see a ghost, your first inclination is always that what you are experiencing is *normal*—not *paranormal*. When you turn around and someone is standing next to you, your first thought is not, "That's a full-bodied apparition." It's, "That's a person." That's how your brain reacts instinctively. Speaking from years of experience, nine times out of ten you won't realize you are looking at a ghost until it disappears—even if you are in the middle of a paranormal investigation. That's what happened to me that night in the Garden District. It happens even now on investigations.

But when I thought it through, I realized that the woman's clothing and hairstyle were better suited to the nineteenth century than the twenty-first. Red was a popular color for courtesans. Could she have been the ghost of one of the prostitutes who worked in the bordello? Possibly a victim of the fire? A madam? I saw her features clearly, but unfortunately there were no photographs on file of the women who worked in the brothel, so there was no way to say for sure. It's possible, too, that the ghost was a former resident unrelated to the brothel. We never found out.

The experience was intriguing, but by far the most fascinating bit of evidence turned up the next morning as we combed through our audio recordings. Upstairs in the bathroom in Kevin's condo, we caught a Class A EVP of a man's voice just after I asked the question, "What is your name?"

"Name's Riddell," it proclaimed.

Could it be a ghost attempting to identify himself?

When we got home to San Antonio, we threw ourselves into research on the history of the Garden District and the street where the house we investigated was located. We learned that, in addition to the bordello, the site where the condo now stood had housed an apothecary (or pharmacy) in the nineteenth century. To our amazement, we also found a prominent New Orleans resident named John Leonard Riddell.

Riddell was born in 1807 in Massachusetts, and enjoyed a long, impressive career as a botanist, geologist, doctor, chemist, and lecturer on scientific matters, as well as a science fiction author. From 1836 until his death in 1865, he was a professor of chemistry at what was then the Medical College of Louisiana, now Tulane University, in New Orleans. He also worked for a time as a melter and refiner at the historic New Orleans Mint, which produced more than 427 million gold and silver coins between 1838 and 1861, and again between 1879 and 1909. The landmarked building, at the edge of the French Quarter, is the oldest surviving structure to have served as a U.S. Mint.

Could Riddell be leaving evidence of his trade even now, in the form of the antique coins Kevin kept finding? Apothecaries dispensed medicine and general medical advice in past centuries. Had Riddell, an MD, spent time in the shop that once stood on this site? Might he have had an emotional connection to the location?

He died the year the Civil War ended. Could the Confederate officer whose ghost appeared in the halls have been one of Riddell's friends or acquaintances?

Riddell fascinated us for a number of reasons. He conducted one of the earliest and most extensive American microscopic investigations of cholera; he believed that electricity held important answers to the quandaries of modern science and that it held the key to curing medical problems; and he lectured widely to help bring a better understanding of science to the general public. He also ran for governor of Louisiana and published a science fiction story about a man who travels to the moon and Mars.

He was an out-of-the-box scientific thinker, and that struck a chord with us. He went to Texas repeatedly for scientific research, which meant he had spent time on our home turf. If we felt some sort of connection to him, could he have sensed the same in us and responded to our presence when we investigated the house? We've conducted experiments that suggest to us that ghosts can tap into strong human emotions like fear and use the energy they produce to manifest. We've also done investigations that indicate certain entities are more likely to respond to specific individuals than to others. For example, in *Ghost Lab* Season One, we locked our female team member Katie Burr in the solitary confinement cells of Alcatraz alone, theorizing that the spirits of men once trapped here without female companionship might retain either the intense longing for a woman's company that they felt in life or possibly deep-seated rage and resentment toward women. Sure enough, Katie's presence triggered a hair-raising EVP that hissed, "I want you!"

Katie became an Everyday Paranormal member more or less by accident. She was Mike Berger's girlfriend when we met her and she showed up at her first investigation planning to be a bystander. But it was obvious she had an aptitude for paranormal work and before long she became invaluable. She wasn't afraid of hard work. She never shrank from a challenge even when she was secretly scared to death, and having a female presence often generated paranormal responses male ones didn't.

We wondered how much a ghost could "read" about a living human when it came in contact with one. If it could sense certain information about you—whether simple facts like your age and gender or more complex ones about your past or your thoughts—would an entity be more inclined to respond to you or to interact with you? To borrow a cliché, would it sense a kindred spirit in the room?

Then again, even if this was an intelligent entity answering our questions, it could have been a different Riddell—someone whose name isn't remembered in the history books. Or the ghost might have been trying to provide different information—Riddell could have been the name of someone significant to him, for example. If there had been a crime on the site, the phrase could have been a clue to a killer's identity. And the hypothetical connection we made to the mint might be incorrect. Kevin showed us the antique coins, but we had no way of proving definitively that they materialized out of thin air. We trusted our client and her friend, but there might be a simpler explanation. One of them might have made up the story about the coins.

Still, the EVP was a milestone for us at Everyday Paranormal. Not only had we gotten a direct response to a question we had asked of a paranormal entity, but it was also the first time we had successfully matched a name heard on an EVP to solid documented historical records. (If you've watched *Ghost Lab,* you'll know that we had a similar success in Gettysburg at the beginning of Season Two, when we asked an entity in the historic Fairfield Inn what its name was and were told "Private John Riley." With the help of a local historian, we uncovered data showing that there was indeed a Private John Riley who died during the Battle of Gettysburg and whose body was never found.)

We visited Maxine again to conduct a follow-up investigation on the way back from filming an episode of *Ghost Lab* at the Myrtles, our third trip to the plantation. Investigation number two yielded an EVP of a woman singing that we thought might correlate to my sighting of the woman in red, though we couldn't back it up with data. We also gave Mike a chance to conquer his fear of the crawl space by sending him back into it again, this time followed by another lithe and lean production crew member. Ten team members stood by to shine extra light down into the hole and to be ready to pull the pair of them out if anybody panicked.

This time, they managed to crawl all the way to a long-disused and crumbling subterranean door and take photographs of it. It did look a lot like the nineteenth-century crypts you see in New Orleans cemeteries. Unfortunately, though, they could find no name or inscription on it. We never managed to figure out who or what was lying beneath Maxine's house. In all likelihood, it *was* a grave and, like countless others, progress paved over it and obliterated the names of the people laid to rest there. The last we heard, the foundation work had been completed and nobody was complaining about any menacing entities under the floorboards, although the friendlier ghosts on the upper floors continued to make occasional appearances.

10
Whispers in the Walls

BRAD: Laurel* was living alone in a rented house on the outskirts of Fredericksburg, Texas, when she called us for help in the winter of 2008. There was definitely something strange going on, she told us. At first, the weird little incidents had just puzzled her, but lately they had escalated to the point that she was becoming genuinely terrified of the place. "It's getting so bad that I'm thinking about breaking the lease and moving," she confessed.

On several occasions, Laurel had draped sheets over the flowers in front of her house before going to bed at night to protect them from the cold air. Each time, when she woke up the next morning, she found the sheets neatly folded and placed on her porch. She checked with the landlord to make sure he hadn't stopped by and removed the sheets. The house was isolated, so she couldn't dismiss it as the work of busybody neighbors. Maybe some kids had passed by on bikes and decided to play a prank on her?

Then late one evening about a week later, somebody rang her doorbell. Laurel walked over to the door and, rather than open it, she peeked out the glass pane at the top. There was no one standing outside. The next night, the same thing happened. Before long the bell was sounding every night. Each time it did, Laurel would trudge wearily to the door and peer out the pane at the top, only to find that the front porch was deserted. The routine was annoying and unnerving for a woman living alone in such a remote location, but she still thought it was someone's idea of a practical joke. She assumed she was the victim in a ding-dong-ditch-'em game, and a couple of preteens were probably crouching behind the bushes nearby, cracking up whenever they saw her face at the door. She thought about calling the police to report what seemed like harassment, but after what happened next, she contacted us instead.

It was around midnight and Laurel was sitting alone in her living room watching TV when she suddenly felt overcome by the eerie sensation that somebody was watching her. Thinking instantly about the pranksters she assumed had been ringing her doorbell every night, she glanced toward the window. To her horror, she saw the face of what looked like a Native American man looking in through the glass at her. Too shocked and frightened to move, she stayed rooted in her chair, her eyes locked on his. As she stared at him, his features slowly morphed into the face of a leering white skull against a black backdrop.

She raced to her bedroom, locked the door, and called the police. But when they showed up, they couldn't find any sign of intruders. There weren't even any telltale footprints outside the window.

Laurel wasn't surprised. She was sure that the terrifying face belonged to something supernatural. There was no doubt in her mind that the house she was renting was haunted—and not by friendly spirits. Given the location, it made sense. The property was in the middle of what had once been Apache and Comanche territory, and it bordered Enchanted Rock, a gigantic pink granite dome where archaeological research indicates the ancestors of the Plains Indians lived more than 10,000 years ago. As the name suggests, Native American legend held that the dome was a highly magical place. Some believed it was a gateway to the spirit world.

We agreed to check out Laurel's house, a tiny 1,200-square-foot ranch, to see whether we could collect any evidence to substantiate or allay her fears. Of course, there were a number of possible explanations that had nothing to do with ghosts. Laurel might really be the victim of a prankster or a peeping Tom who had decided to take his game to a new, more threatening level. You can find battery-powered Halloween masks that change color and that let you create other special effects. Maybe somebody in a mask stood outside Laurel's window at midnight just to scare her.

Then again, Laurel might have had what's called a hypnagogic hallucination, a sort of waking dream that happens when a person is hovering between sleep and wakefulness. The episode can even be accompanied by temporary full-body paralysis. People don't always recognize it as a product of their subconscious mind because they are convinced that they were still awake when it happened.

We decided to bring along a member of the newest unit of the Everyday Paranormal team—the K-9 Unit—when we visited Laurel's house. (Naturally, we got permission from her and from the landlord first.) The concept of using animals in paranormal investigative work would catch on soon and inspire a TV series a year later, but when we headed out to Fredericksburg that cold winter evening we didn't know of anyone else who was experimenting with the technique. We just recognized that animals, dogs in particular, had more keenly tuned senses than humans and we thought it would be interesting to observe them during our investigations to see whether their actions corresponded to the evidence we collected. If they did, we would have another way to correlate data. The dogs we used weren't specially trained. They were just our own everyday pets and those of our team members.

Incidentally, we weren't disappointed by the K-9 Unit experiments. In a number of investigations, our dogs reinforced what we caught on video and audio. They reacted to the rooms where we caught evidence, and they were most panicked in the cases where the tones of the EVPs were most menacing. Why did it happen? Are animals psychic? You hear all kinds of crazy superstitions about animals being able to see ghosts. There's even an old wives' tale that says if you stand behind a dog when he sees a ghost and look over his head directly between his ears, you'll see the ghost, too. That's nonsense, of course. We think there's a more logical, scientific explanation.

Animals are tuned in to things humans miss altogether. For example, lots of them know when a storm is approaching and react, probably because they are more highly sensitized to electromagnetic activity and subtle changes in the environment like shifts in air pressure. That might explain the numerous cases of animals anticipating earthquakes and other natural disasters. It's not intuition; it's biology. It made sense to us to hypothesize that dogs might react to subtle changes in energy level and other environmental factors triggered by paranormal activity, too.

Some dogs know when their owner is coming home ten minutes or more before he reaches the house. They start hovering around the front door, whining, pacing, and thumping their tails long before they hear footsteps or the rumble of tires in the driveway. Why do they do that? Certain domestic animals can also tell when a human is happy, sad, frightened, angry, or sick. And they respond accordingly, staying out of the way if you are in a bad mood or hopping into your lap to nuzzle you if you are depressed. We didn't think that had anything to do with telepathy. But we thought it might suggest that animals have the ability to pick up on the energy certain strong emotions generate. It stood to reason that if ghosts generated anything resembling human emotions like grief or rage, a dog might pick up on the energy shifts those emotions generated, too, and exhibit some sort of corresponding action.

The dog that accompanied us to Laurel's house was a five-year-old male Boxer named Kito, with a friendly, mellow disposition. He came into the place wagging his tail, and we led him through the various rooms, letting him sniff and explore to his heart's content. He showed his usual amiable curiosity in most parts of the house. But then, to our surprise, as he neared the master bedroom, he froze. You could see his entire body stiffen and his hackles rise. The fur along his neck and backbone bristled. A split second later, he spun around and tore down the hall away from the bedroom as if he had gotten zapped with an electric shock. The animal was beyond terrified. He slammed headlong into a wall because he was so desperate to get away. The pain of banging his head didn't even seem to faze him. He just bounced off the wall, scrambled to regain his footing, and lit out of there.

We marked the time and location of the dog's extreme reaction and continued the investigation. Back with the rest of the team in the living room, Kito calmed down, though he refused to venture anywhere near the bedroom again. We went back to the bedroom without him and launched into another EVP session. Barry made an offhand comment that struck me as funny and I started chuckling. He joined in and, before long, we were both laughing uncontrollably. If you ever saw us on *Ghost Lab*, you'll know that we've both got a sense of humor and that we are prone to crack up at times when you would expect us to be serious. Sometimes, one of us will crack a joke to relieve the tension of sitting in the dark for hours trying to catch an EVP. At other times, we make jokes as a deliberate technique designed to elicit a paranormal response.

During that session, we were so loud that we wouldn't have heard much anyway, but on the recording the next day a sudden heavy pounding was clearly audible over our laughter, as if someone was banging repeatedly on the bedroom wall in response to the noise we were making. Then, a deep, loud male voice resonated on the tape saying what sounded like the word "Jones!" On the recording, we continued to crack up and talk as if it had never happened.

A short time later there was a distinctive growl and a voice saying matter-of-factly, "Six o'clock!" Perhaps the most disturbing of all was a direct response—one of the first we ever captured—to my provoking. I was getting impatient with the lack of activity and I told the ghost, "You need to get out here right now!" On the recording, the same deep masculine voice that characterized the other EVPs answered, "No!"

All the recordings were Class A EVPs—clearly audible voices speaking easily distinguishable words in English. All came across at a frequency too low to reach the threshold of human hearing and too low to have been actual human voices or random radio broadcasts we might have mistaken for disembodied spirit voices. And every one of them sounded angry and menacing, in our opinion.

As far as ghosts that might be haunting the place, the possibilities were endless. You couldn't find a location more rife with ancient Indian legends and creepy ghost stories than Enchanted Rock. The spirits of Native American and Spanish warriors fallen in battle supposedly roamed the Rock at night. Members of a forgotten tribe slaughtered by rivals were said to wander the area bent on revenge. There were even tales of bloody human sacrifices taking place atop Enchanted Rock and disembodied screams echoing down the hillsides. According to one story, you could still see footprints in the rocky surface made by the ghost of a chief who sacrificed his daughter and was condemned to haunt the barren granite dome for eternity, filled with remorse.

Over time, sensational stories like these tend to accumulate around famous historic landmarks. They're the kind of tidbits tourists and Boy Scouts on camping trips gobble up. But there is often a kernel of truth even in the tallest tale. Was it possible that restless Native American spirits were manifesting at Laurel's house? If so, why was the activity increasing so dramatically? Why was it becoming so much more hostile? There had to have been some change in the environment that was triggering it. Something was serving as a generator, powering up all that paranormal action.

We sat down with Laurel to share the evidence we had collected during our investigation and to probe more deeply into possible explanations for the heavy paranormal activity she had been experiencing recently. Nothing in our initial interview suggested the kind of dramatic change in an environment that would give rise to the sudden appearance of ghosts, whether ancient spirits tied to Enchanted Rock or more recent ones linked somehow to the actual house Laurel was renting.

"You said you've lived here for almost a year, but all this activity just started recently, right?" I asked her.

"That's right," she said. "It's been happening for a little more than a month now."

"Has *anything* changed about your house recently?" I asked. "Have you or your landlord made any renovations or repairs?"

Laurel shook her head.

"Is there any construction going on around here?"

"Not that I know about."

I wracked my brain. There had to be something we were missing. "Have you moved anything outside the house? Maybe dug up some of the dirt to plant flowers?"

"The flowers were there when I moved in," she said. "The only thing I moved was that big pile of rocks in back of the house."

"What rocks?"

"I was going to make a fire pit for barbecuing, so I started moving a bunch of rocks somebody left out back."

Bingo.

Given the history of the place, it was quite possible that those rocks had held significance to some group of Native American people at some point in time. For all we knew, we were sitting on sacred ground from centuries ago. Maybe this was a burial site. A guardian spirit might have been sent to watch over those rocks and the bones that once lay beneath them. If Laurel had accidentally disturbed the final resting place of someone's ancient ancestors or even of someone with an attachment to that stretch of land and the little house sitting on it, she might well have ticked off a ghost or two.

We wanted to find out how moving that pile of rocks had altered the physical conditions. It seemed to us that Laurel had somehow accidentally turned the area into a better conductor of electricity, creating the kind of high EMF that we are convinced is conducive to paranormal manifestations. Granite is known to contain higher-than-average levels of uranium, and houses in areas with high uranium levels are more apt to have high levels of radon in them. Was uranium or the radon it produces when it decays hidden in the soil under the rocks or in the rocks themselves? If so, Laurel's actions might have released it and given ghosts that had wanted to manifest for who knew how long the opportunity at last.

We offered to come back with ground-penetrating radar, vibration-testing machines, and other equipment that might have yielded more data about the geology around the property. We were also willing to look into the history of the house and the land under it to see what we could unearth about former occupants, builders, and what might have stood there before Laurel's house was constructed. But when Laurel heard our EVPs, she made up her mind.

"I didn't want to say anything to y'all, but I had a psychic visit the house before you did," she informed us. "The psychic told me she picked up a male presence in the bedroom and that she heard a lot of hammering, as if the ghost might be a carpenter.

"She also told me she was getting the name *Jonas* or *Jonesy*."

We were speechless. It was the first time that a psychic's insights had ever meshed with our evidence. Laurel's information reminded us that, in our line of work, you've got to stay nonjudgmental and open-minded. Sometimes, legitimate corroborative data can come from unlikely sources.

Unfortunately, we never got a chance to return and dig more deeply into the bizarre goings-on at the rental house in Fredericksburg. Laurel broke her lease, packed up pronto, and moved out a week after we reviewed the evidence with her.

11
The Flying Tape Fiasco

BRAD: We soon realized that we couldn't expect every investigation to produce great evidence like the Harlequin and the Freeman had. Some were total duds. Believe me, we've spent our share of nights hunkering down in the dark only to come up puffy-eyed, sleep-deprived, and—after poring over hours of audio recordings and video footage—empty-handed. Jessica would often stare at me as I slumped over the breakfast table in the morning and shake her head. "Look at you," she would say incredulously. "You're exhausted and now you've got to work all day. Are you sure this is worth the effort?"

"Absolutely," I would respond, stifling a yawn and reassuring her as well as myself. Barry and I never envisioned Everyday Paranormal as a profit-making venture. It was strictly a hobby. As far as hobbies went, it was a pretty grueling one. Fortunately, though, there were enough successful investigations to keep our adrenaline high and to balance out the low points.

While we're on the subject of low points, no story of our early adventures in ghost hunting would be complete without acknowledging a few of them. The night we came closest to hitting rock bottom was the one we spent at WOAI-News 4 in the spring of 2008. The station was the local affiliate of the NBC television network and the granddaddy of San Antonio broadcasting. It aired its first radio broadcast back in 1922 and became the city's first television station in 1949. It got to keep its call letters even after a 1938 government regulation required all stations west of the Mississippi to start with K because it was already such a venerable institution way back then.

The station was still housed in its original downtown headquarters on Navarro Street—in a building that had opened in 1920 as the Embleton Motor Company. One of the city's first car dealerships, Embleton sold Moon Automobiles and later, after WOAI started broadcasting from its second floor, Chryslers. The dealership eventually closed and the station took over the entire premises, gradually sprawling down the block and adding studios and offices as operations expanded.

The Navarro Street building housed an impressive collection of memorabilia from the station's long and illustrious history. And, if the rumors were true, that wasn't the only lingering reminder of bygone days: The place was supposed to be haunted. Some folks thought the ghost was a one-time reporter who hanged himself at the station in the 1940s, though we couldn't find any historical records to substantiate the suicide claim. Staff members said they heard strange noises when they were working alone, as if an invisible someone was moving around nearby. It happened in the memorabilia room, the studios, the halls, and most often in the room where tapes from old broadcasts were stored. Some employees even swore they had seen tapes fly off shelves by themselves as if someone had flung them.

Everyday Paranormal had been lucky enough to be featured

in an article in the local newspaper, the *San Antonio Express-News*, a few weeks earlier and the story caught the eye of a WOAI editor named Joyce*. The idea of a ghost hunt piqued her interest, so she ran it by her boss and got the okay to invite us in to investigate the station. When Joyce called, we were delighted. It was one of the first times a client had contacted us rather than the other way around—and this wasn't just any client, it was an old and well-respected news station. It sounded like a win-win situation: good exposure for us, and, hopefully, a good news story for them.

The WOAI investigation would have to be a first-rate operation. Barry and I planned it out meticulously. We double- and triple-checked our equipment. We asked our team members to supplement it by bringing their own cameras to ensure that we wouldn't miss a shred of potential evidence. Then we loaded up our gear and pulled into the WOAI parking lot around 8 P.M. Joyce greeted us and introduced us to her colleagues on the news crew before taking us on a brief tour of the building. Along the way, we set up DVRs, still cameras, infrared cameras, and audio equipment at strategic points.

We ran a few preliminary EVP sessions during our walk-through. Several of our team members claimed to have mild personal experiences, including a freelancer named Celeste*, a self-proclaimed psychic in her mid-twenties who tagged along with us occasionally at her own request.

Our plan was to get the heavy-duty investigation underway after the eleven o'clock news had wrapped and most of the staff had departed, minimizing the risk of distractions and false readings from broadcasts interfering with our equipment. We knew there was a lot of electronic equipment in the building, so we paid particularly close attention to baseline readings. In the meantime, Joyce invited us to watch the news live as it aired. We followed her to the studio while the rest of our team continued investigating, running EVP sessions, and exploring the building.

The weather report was just ending when Steve Harris came bolting in. "Man, you've got to come and check this out," he exclaimed breathlessly. "You won't believe it!"

As we raced after him, Steve explained. "We were in the tape room, running a session, and one of those tapes literally came sailing off a shelf. It almost hit Celeste in the head!"

Hearts pounding, we reached the archive room with its towering stacks of shelves. Sure enough, toward the back of the room we spied an empty spot in the neat rows of black plastic cases on a shelf about five feet high. On the floor a few feet away lay the missing tape. We did a quick calculation of the distance and the trajectory. We picked up the tape to see how heavy it was and estimated how much force it would take to make a tape of that weight move that far, at that angle, from that height. Could gravity have made it fall? Doubtful. Could it have been put back carelessly so that it was balanced on the edge of the shelf? Again, unlikely. We had walked through this room about an hour earlier and inspected the shelves fairly carefully. It was hard to believe that not one of us would have noticed a tape out of place. Could a team member have accidentally knocked it down? Possibly. But Steve said the team had finished the session and was leaving the room when the tape moved. In fact, Celeste, the last team member to exit, claimed the incident had taken her completely by surprise. She hadn't even gotten her usual psychic "vibes" to warn her that a spirit was hovering nearby.

Aware that the room was supposed to be a hub of paranormal activity, we had positioned one of our DVR cameras atop a shelving unit earlier in the evening and left it running. "Man, if this really happened, we caught it on video," Brad whispered to me, trying to contain his excitement. This might turn out to be some of the most memorable footage we had ever captured. The possibility was so intriguing that we conferred for a few minutes and decided to break standard protocol: For the first time ever, we would review our footage in front of a client.

"We *never* do this," I told the news crew. "But we have to see what we've got on tape here."

BARRY: We retrieved the DVR and led the way to the office we had established as our command center. Everyone from our team crowded in eagerly along with Joyce and a WOAI cameraman friend of hers who was intrigued by what we were doing.

An image of the tape room, viewed from the camera's position high up on the shelves, flickered onto the playback screen. It showed our group of investigators walking in and initiating the EVP session. You could hear Steve introducing himself and running through the usual prompts. "Is there anyone here? Did you work here? When?" The rest of the group stayed near him, but for some reason Celeste wandered off toward the back of the room, near the shelving unit that held the falling tape. Was she genuinely psychic? Had she sensed a presence in that row or possibly even drawn a ghost to it?

The tape showed the session continuing and then finally ending. The team gathered their personal equipment and headed out the door. Celeste trailed behind the others, dawdling. As she passed the shelves, she stretched a hand up tentatively toward a row of tapes. *What the hell was she doing?* She cast a furtive glance toward the door, then reached out and hooked one of her fingers lightly over a tape. It slid a fraction of an inch toward the edge of the shelf.

"Stop the tape," I thought desperately. "Stop the friggin' tape!" But there was no way to reach the switch in time. Besides, unless I was very much mistaken, everyone else knew what was about to happen just as well as I did.

Celeste repeated the movement, reaching farther over the tape and tugging harder. This time her actions produced the desired result. The tape flew past her off the shelf and landed with a thud on the ground.

"The tape!" she cried, gazing toward the rest of the team in the doorway in mock amazement. "It scared me!"

Nobody in the room said a word as Brad hit the Off button. Every head swiveled toward Celeste.

Realizing we were all staring at her, she cleared her throat. "What happened?" she ventured in a small voice. "Did I bump it?"

We're finished, I thought. That's the end of Everyday Paranormal right there on that tape. All that work. All those years of reading and research. All because some idiotic woman we hardly knew and hadn't wanted to bring along anyway decided to falsify evidence and was too stupid to realize she had done it right in front of a video camera.

And this had happened not in some innocuous, obscure spot but at a news station, of all places, with people from the news station's staff watching it! Odds were that it would make the six o'clock news the next day. If I were a newscaster, I would run the story. It was almost too good to resist. We would be the laughingstock of San Antonio in less than twenty-four hours. We would set paranormal investigation back by years, discredit the entire field, when our mission from the beginning had been to do the opposite.

I didn't meet anyone's gaze. I just stood up and started packing our bags. The rest of the team followed my lead. Meanwhile, Brad pulled Joyce and the cameraman into another room to do damage control. I could hear him apologizing profusely. "This is *not* the way we do business. It's never happened before and it will *never* happen again," he assured them. "She is not a regular member of our team and she will never work with us again.

"That's because I'll probably kill her if I see her," he added, only half kidding. At least he could muster some humor. The rest of us were too stunned and furious to speak.

To my amazement, Joyce accepted his apology graciously. "Don't worry about it," she said. "It's no big deal."

We watched the news for days, paralyzed with fear. It was scarier than any ghost investigation we had conducted. But, miraculously, the station never broadcast the gaffe.

As for Celeste, she vanished like a ghost herself. By the time we dragged ourselves back out to the parking lot that night, disgraced and defeated, she was nowhere to be found.

I called her repeatedly the next day, but she refused to pick up the phone. "Well, you know why I'm calling," I finally told her answering machine. "This is your official notice. You are no longer a part of the Everyday Paranormal team."

* * *

BRAD: About a month went by and then, to my amazement, Celeste called. "Hi," she said cheerily, as if nothing had happened. "I was just wondering when your next investigation is."

"You're kidding, right?" I asked, flabbergasted.

"No. I was thinking it would be fun to come along."

"You've got a hell of a lot of nerve," I told her. "That stunt of yours at WOAI could have seriously damaged our reputation. You could have put us out of business."

Now it was her turn to act flabbergasted. "I don't understand," she said. "What did I do wrong?"

"You faked evidence! You yanked a tape off a shelf. On camera! Don't you remember?"

There was silence on the other end of the line.

"Listen, Celeste. Don't call us anymore. Have a nice life."

Psychosis? Denial? Impulse control issues? Those aren't subjects we investigate. We don't want to explore the disturbed psyches or secret motivations of the living—just the dead. Was she deliberately trying to discredit us? We think it more likely that she just fervently wanted people to believe she really was psychic, a lightning rod for paranormal activity, so she decided to go to unethical lengths to convince people of that. Needless to say, her actions at the news station made us question every EVP and picture she had caught in her earlier work for us. Just to be safe, we decided to scrap all the evidence she had collected. Fortunately, there wasn't much of it.

We could have posted the footage of her on YouTube. We could have dragged her name through the mud. At one time we were almost angry enough to think about it. But that's not our style. What would be the point? It wouldn't advance the field of paranormal research.

So instead, we licked our wounds and moved on. Celeste wouldn't be the last fraud or liability we would encounter, but she taught us to be wary, to look for danger signs like inflated egos and lack of scruples in potential team members, and to try our best to surround ourselves only with people we trust implicitly.

Incidentally, we still have the footage of Celeste and the infamous flying tape. But when we watch it now, it's hilarious. Sometimes you've got to get a little distance from a situation to see the humor in it.

After all that, we reviewed our tapes the next day and were surprised to unearth a legitimate EVP of a male voice that said, "Score the tape." We weren't sure what it meant or whether we were interpreting his words correctly until we consulted some friends of ours in broadcasting and they told us that it's a common editing term. We double-checked the frequencies to ensure that the phrase wasn't just a scrap of conversation from the news station that our recorders had inadvertently picked up, but the frequency was far too low for a human voice. The recording carried all the telltale characteristics of a disembodied spirit voice. However, we were too chagrined to contact the news station in the wake of the Celeste debacle. Even under different circumstances, we would consider a solitary EVP scant evidence, so we simply filed it away in our archives, where it remains to this day.

12. The Ghost in the Bedroom

BARRY: In September 2008, Everyday Paranormal got a call from a woman named Carol Ecker*. From the moment the conversation started, I could tell she was stressed and edgy. "I heard about you from a friend," she began. "I think my house is haunted. My whole family is scared. Can you help?"

The house in question was a rental in a small town not far from our home base. Carol had been living in it for two years with her husband and two children from a previous marriage—a twelve-year-old son named Charlie* and a sixteen-year-old daughter named Margaret*, who was a junior at the high school across the street from the house.

"We've thought the place was haunted pretty much ever since we moved in," Carol explained to me. It had started with small objects going missing and then showing up in different rooms when no one in the family had touched them. Then the family began hearing footsteps—heavy thuds pounding across the second floor. Jim Ecker* would rush upstairs repeatedly, afraid an intruder had broken in. But the second floor was always empty.

The Eckers weren't superstitious people, not the type to

believe in ghosts. They thought there might be some obscure but logical explanation for what was happening in their home. But before long, the events took a more sinister turn. At one time or another, each member of the family felt an inexplicable sensation of being shoved hard from behind as they walked up or down the stairs, though there was no one near them. When they started to see full-bodied apparitions of a young man standing in various parts of the house—a stern, ominous figure silently watching them—it left them with little doubt that something inhuman was living in their midst and growing increasingly hostile toward them.

"It's getting worse," Carol told me, working to control the panic in her voice. "We're starting to get really worried about Margaret." The spirit, it seemed, had fixated on the teenage girl and was tormenting her relentlessly.

For months, Margaret had been hearing the ghost whispering to her in increasingly urgent pleas. Now it had turned threatening. It was hurting Margaret, sometimes even choking her, Carol claimed. The girl was so terrified that she had refused to sleep in her bedroom for more than a month. Now she wouldn't even set foot in the room alone. When she had to get a change of clothes in the morning, she begged her brother to come with her and stand in the doorway for protection.

Margaret was hardly your average kid. She was a straight-A student and an outstanding athlete, sure to get a full scholarship to college and likely to make it big someday. But her mother was convinced that whatever was haunting the Eckers' house was destroying her. The girl was sleep-deprived, antsy, and incapable of concentrating on anything. Her schoolwork was slipping. In her mom's opinion, she was close to the breaking point.

A destructive spirit terrorizing a brilliant teenage girl and wreaking havoc in a nice suburban home? The story hooked us right away. The activity the Eckers described wasn't unusual. As you know by now, we hear tales of mysterious footsteps, moving objects, and even full-bodied apparitions pretty regularly. But it was extremely rare to hear about an entity targeting a specific person and trying to hurt them. A push or prod isn't necessarily a hostile or warning gesture from a ghost. It might just be an attempt to interact. But getting choked? We had never encountered that magnitude of hostility or intense normal-paranormal interaction before. If what Carol told me was true, there was an unusual urgency to investigate and find the real root of the Eckers' problem.

Of course, there were a lot of potential explanations that had nothing to do with paranormal activity. The Eckers might be making the whole thing up; they wouldn't be the first. They might have overactive imaginations. They might want attention. They might just think hanging out with a bunch of ghost catchers for a night sounded like fun. They might even be noticing the first signs that their overachieving teenager was crumbling under intense pressure. Margaret's parents might be grasping at any straw but the obvious to explain away their daughter's increasingly fragile mental state.

There was no way to know until we investigated. Carol wanted us out there asap. "Friday night is homecoming at Margaret's school," she said eagerly. "We'll all be out late. We can even stay with our friends next door, if it'll help. You can have the place to yourselves. Will you come?"

We agreed and, as the sun was setting and the weekend beginning, the Everyday Paranormal crew loaded up our trucks and drove out to the Eckers'. The setting was nondescript—your average two-story suburban house in a tract of nearly identical houses, all built by the same developer sometime in the 1990s. It was hardly the foreboding moss-covered mansion you see in the movies, but experience had long since taught us never to judge the book by the cover. A building's facade tells you nothing about its potential for paranormal activity. If anything was haunting this place, it might be a holdover from people who lived and died here or traumatic events that unfolded here long before the site was bulldozed.

In fact, I knew firsthand that this area had seen grisly deeds before its transformation into a subdivision. Back in my days as a paramedic, this was lonely, desolate farmland. I could remember responding to a call here late one night and finding the blood-soaked body of a decapitated teenage boy. We later learned that he had crossed the Mexican mafia and paid for the mistake with his life. These streets had changed a lot, but driving through them at night still brought images of that gory headless corpse flooding back through my mind.

I tried to shake it off. We need to go into all our investigations levelheaded, open-minded, and objective. And the power of suggestion can be a dangerous thing. Everyday Paranormal's mission is not to validate a client's claims or to prove them wrong. It's to conduct a thorough, neutral investigation and then to examine the evidence and draw conclusions based only on that.

Steve and Jason were with us on the Ecker house investigation. We were still experimenting with the K-9 Unit, so we also brought along Brad's Chihuahua, Cocoa. It would be Cocoa's first investigation.

We pulled into the drive and started unloading our equipment. As we tromped into the living room, recording devices in tow, Carol welcomed us gratefully and introduced us to her family. Then they hurried off to catch the homecoming game kick-off, promising they would stop by later to find out how the investigation was coming along.

We set up an infrared DVR at the bottom of the stairs facing up toward the second floor, where the apparition reportedly materialized most often, as well as upstairs, both inside and outside Margaret's bedroom, the epicenter of the paranormal activity, if the family's information was accurate. Steve was stationed in the living room where he would have a clear view of the staircase, and Jason took up a post in the dining room to continuously monitor the DVRs. Brad and I planned to conduct EVP sessions, starting upstairs.

Per our standard protocol, we turned off all the lights. Residential neighborhoods rarely get pitch-black, thanks to the abundance of porch lights, neighbors' lamps and TVs, and so on. With the glow of the streetlights and the other external light sources nearby, infrared gave us plenty of light to cover the house effectively. We took baseline readings of the temperature and humidity and then set up the data loggers we had added to our arsenal in a line along the upper floor, leading into Margaret's room to create a linear sweep that would alert us to any notable changes in the environment, which might signal that an entity was passing through. Data loggers gauge not only temperature but humidity and dew point as well—handy if you encounter ectoplasmic mist or entities that don't just alter the ambient temperature but increase the dampness in certain spots.

One of our first moves during the investigation was to take the dog upstairs. Brad led with Cocoa, and I followed. We had gotten about halfway up the steps when I felt a sudden forceful shove against my back as if someone was trying to knock me down. I whirled around angrily to see whether Jason or Steve was behind me, playing a joke. (Hey, it happens once in a while.) There was no one there. The sensation was just like the one Carol Ecker had described—aggressive and unfriendly. We chalked it up as an intriguing personal experience, but no more.

What happened next gave us more solid data suggesting that the house really was haunted. Cocoa is one of the most easygoing canines any of us has ever met. He loves everyone. But the second that dog got near Margaret's bedroom, he started whining and trembling. Within seconds, he was struggling desperately to get back down the hall, clawing at his own collar to get away. Brad literally could not drag that tiny Chihuahua into Margaret's bedroom. When he tried, Cocoa growled and snapped at him. It was totally uncharacteristic. Finally, he gave up and let go of the leash. Cocoa bounded back down the stairs, where he hopped happily up onto the couch next to Steve and started wagging his tail again. We repeated the experiment several times throughout the night to make sure the dog's initial reaction wasn't a fluke. Sure enough, each time he got near that bedroom, Cocoa turned into a beast.

As the night wore on, we collected more and more evidence to support the Eckers' claims of paranormal activity. We were sitting in the living room around 10 P.M. when a fork suddenly flew off the table and hit the floor. None of us had touched it. No one had turned on a powerful fan or put a large magnet near it. We could come up with no logical scientific way to explain the occurrence away.

Shortly after that, loud noises sounded over our heads, as if someone or something heavy was running from one end of the second floor to the other at full speed. Was somebody up there racing around without our knowledge? Brad and I ran up to see, but the second floor was deserted.

Just about that time, the family walked in the front door, curious to see how our investigation was progressing. No sooner had they stepped in than the noise overhead started up again.

Everyone was bug-eyed. These were only eight-foot ceilings, and the sounds were booming enough to rattle the rafters. It sounded like a man in heavy shoes was running back and forth.

There were eight of us standing in the living room together, and every one of us heard it. No way could this be dismissed as the product of anyone's overactive imagination. The ceiling was literally shaking.

We counted heads to make sure everyone was accounted for. Then Brad ran upstairs once again. He checked every room, every closet on the second floor, but found no one. It was a fantastic piece of evidence, not only because we caught it on both audio and video, but also because a large group of witnesses could verify it.

The evidence we collected the next day in the lab turned out to be equally convincing: We painstakingly replayed the hours of tape we had recorded, straining our ears for EVPs. At one point on the tape I heard my own voice asking, "Where are you?"

Though I didn't hear an answer at the time, there was a clear response on the tape.

A man's voice said, "I'm right here next to you."

Believe me, the words weren't spoken in a warm, amiable tone. There was menace in the voice.

We caught an equally unnerving EVP when we conducted a follow-up investigation on a Tuesday night a few weeks later.

One of the questions Brad asked during the second investigation was, "What day of the week is it?"

On the tape there is a clear EVP response in the same deep, unfriendly male voice: "Tuesday."

Even we were impressed. For us, evidence like this is gold. But for our clients, we knew the chilling audio recordings would only inspire more terror.

So we sat down with them to discuss our findings the following afternoon. "Our research indicates that there is definitely paranormal activity in your house," I informed the Eckers. "We had personal experiences, we got a reaction from the K-9 Unit, and we caught EVPs."

The recorded voice, the forceful shove against my back, and the family's own experiences strongly suggested that this was an intelligent entity—something that was aware of the living humans around it and was interacting with them intentionally.

"Since you don't want to move out and neither does whatever is living here with you, you're going to have to take charge," I told the Eckers, directing my comments mostly to Margaret. "You can't let this thing run your life or ruin it."

Brad and I gave Margaret our best advice on how to confront the presence and tell it to leave her alone. She seemed as if she was listening attentively and she assured us she would do what we suggested.

As parents of young daughters ourselves, we both thought the smartest action would be for the Ecker family to move. But that never happened. Even though all the evidence pointed to this being a malevolent spirit and the Eckers all claimed they were frightened out of their wits by it, more than one of Carol's comments made me wonder if the idea of living in a haunted house appealed to her on some level despite the danger signs. She mentioned, for example, that she had told all her friends about the haunting—and the fact that paranormal experts were coming in to spend the night in her house. In our experience, people who do that are usually excited about having contact with a ghost. Whatever the Eckers' real reason, they decided to tough it out with the ghost.

Normally, we don't have much contact with clients after our investigation concludes. But in this case we were concerned about Margaret, and since Carol worked nearby, I checked in with her every so often. I was relieved and happy to hear that Carol's daughter was sleeping in her room again and that she seemed to be coping with the ghost that had become obsessed with her. Brad and I were even happier when we found out that Margaret had graduated early, as valedictorian of her class, and gotten a full athletic scholarship to a prestigious university. She was truly an amazing kid. We both expected to be reading about her accomplishments in the newspapers a few years down the road.

So it came as a shock when I got a phone call from Carol about six months after Margaret graduated.

It was after 11:30 P.M. when the phone rang.

"Barry? It's Carol Ecker," the caller said, her voice trembling. Fighting tears, she told me that Margaret had come home from college for a visit a few days earlier. Within twenty-four hours, they had found her huddled in the corner of her bedroom, shaking and mumbling.

"I'm the only one he thinks is pretty . . . I'm the only one he can talk to . . ."

Unable to coax her out of near-hysteria, the Eckers' family physician had finally admitted the teenager to the psychiatric ward of a nearby hospital. Margaret arrived so pale and trembling so violently that one of the nurses who knew nothing of the family's history remarked, "That girl looks like she's seen a ghost."

At long last, Carol said, the Ecker family was moving out of their haunted house.

"About time," I thought, though I resisted the urge to say it aloud. Instead, I offered Carol my sympathy and told her I hoped Margaret would recover soon.

"Isn't there anything you can do to help us?" Carol asked, her voice pleading.

"I'm really sorry," I replied. "All we do is investigate and find evidence. I can't get rid of it for you."

Had Margaret finally cracked under the pressure of being such an ultra-high-achiever? Was there something damaged in her psyche all along? Had she used ghost stories as a way to hide emotional distress that sprang from other sources? We later learned that Carol's first husband had been abusive—and that the kids had been involved, though we never knew the details. Was it possible that Margaret's breakdown stemmed from something she had experienced in the past? We're not psychiatrists, so we don't delve into those areas. And we don't like to speculate because it doesn't give us any hard data. But we do know that our evidence indicated there was something seriously wrong with that bedroom.

An outsider hearing about the Ecker case might dismiss Margaret's problems as pure emotional instability, stemming from academic stress or abuse. They might even argue that the family's eagerness to believe in—and blame—ghosts sprang from their own guilt or inability to face the real causes.

But that doesn't explain the voice on our tapes, the dog's panicked reaction, or the booming disembodied footsteps that more than half a dozen sane and rational adults heard in the Eckers' house that night.

Had the spirit of an unidentified young man lingering in Margaret's room tormented her so much that it finally drove her over the edge? Based on our own findings, we were inclined to believe the answer was *yes*. Had it fixated on her because something in its own past drew it to pretty young women? Had it chosen her because she was emotionally fragile and somehow her mental state generated the right kind of energy for a ghost to latch onto? The ghost didn't explain its motivations—or at least we didn't capture them on an EVP—so we'll never know.

Another team of ghost catchers might have openly urged the family to move. They might have brought in a priest or a minister to cleanse the house. They might have invited a psychic or a white witch to commune with the presence, burn herbs or wave rosemary sprigs, to find out what it wanted and then help it resolve the issues prompting the angry outbursts so it could "move on."

But as you know, we have never put much stock in those ideas. We've seen it tried; we've never seen it work. We mean no disrespect to anyone else's views, but in our years of studying paranormal activity, we have never been able to collect any data indicating that cleansings or coaxing ghosts to "go to the light" is effective. Then again, the paranormal field evolves every day. If we find scientific data suggesting that exorcisms and communing with spirits produces results, we will be as interested as anyone else in our line of work. But so far, our evidence has shown time and again that if a ghost wants to be someplace—and if it can find energy to attach itself to—it is going to be there.

Since the day we founded Everyday Paranormal, we have had a straightforward mission: to investigate places alleged to be haunted using quantifiable scientific measurements and, based on our findings, to either validate or reject claims about paranormal activity. Some of our clients get peace of mind knowing that what they feared was a haunting is really "all just in their head." Others feel better knowing it isn't.

We weren't sure what was triggering the haunting at the Eckers' house. If we could have isolated a source that was generating high EMF in the area, in theory we might have been able to adjust factors in the environment that would have drained the ghost's power source.

Then again, the energy might have been coming from human sources. When you're frightened, your hypothalamus sets off alarm bells all over your body—your heart rate surges, your blood pressure rises, you start to sweat, and you release a stampede of stress hormones. Our experiments suggest that the strong emotion of human terror and the physical reaction it creates might produce an energy source that acts as a battery for ghosts. If you take that element out of the physical environment, you might be able to cut off the ghost's energy supply and keep it from making its presence known. On the other hand, the ghost in the Eckers' house was active and aggressive when we encountered it, and Brad and I weren't scared. Had it drawn enough reserve energy from Margaret's fear to manifest even when she wasn't around? Or, was there another energy source we overlooked? Ghosts' energy sources aren't always obvious or easy to pinpoint. If they were, ghosts would be all over electronics stores and apartment complexes where everyone has computers and TVs running simultaneously.

Carol Ecker is far from the only client to say, "Help me!" and to ask us, "How do I get rid of this ghost?!" Aside from locating the energy source an entity is using and then taking it away, our research suggests that it's not possible to end a haunting, especially in cases of intelligent energy. If you're not willing to vacate and let the ghost have the space, your best bet is to take charge, as we advised Margaret. Learn to deal with it so you can live with it. If you had a roommate who was bothering you but wouldn't move out, you would probably tell him or her, "Look, you can live here. You can stay, but you can't bother me." If this thing is intelligent enough to interact with you and you are assertive enough with it, it is going to understand that you are in charge and adapt its behavior.

If you can't do that, you might as well get out and let the ghost have the place.

The Ecker case was the only one we have ever seen where a person was being repeatedly emotionally tormented and physically harmed by a ghost. Our feeling was that whether you believe in ghosts or not, whether you think a family member's stories about being terrorized are true or imagined, if you are having that many problems, *leave*. Though ghosts do sometimes follow people from location to location, there is a good chance that whatever has been bothering you will remain in the house after you vacate the premises. And even if you ultimately conclude that your problem was *not* paranormal, the change of scenery just might help you examine, identify, and hopefully deal with the real source of your issues.

13

Ancient Artifacts and Disembodied Voices

BARRY: In the spring of 2008, I took my special education class on a field trip to the Institute of Texan Cultures in San Antonio. Originally constructed as part of the 1968 HemisFair celebration—the first officially designated world's fair held in the southwestern United States—the Institute is essentially an enormous museum affiliated with the University of Texas at San Antonio—65,000 square feet of exhibits in a sprawling 182,000-square-foot complex. Its vast permanent collection and special exhibitions showcase the history of the many ethnic groups that have played a role in shaping the history, heritage, and culture of the Lone Star state.

Among ghost enthusiasts, it is also famous for being haunted. There are dozens of stories about strange, inexplicable events occurring among the artifacts over the building's forty-year-plus history. One former director was a pipe smoker and, though he died years ago, employees still claim to catch the familiar sweet, heavy scent of his tobacco smoke hanging in the air, especially when they work late at night. An elaborate antique funeral coach stands in the center of one collection, and security guards often walk into the room after the Institute has closed for the night and find the coach's back doors where the coffin would be inserted wide open. They close them and move on to their rounds of other rooms and floors. Hours later, they cross back through the room and find the doors open again. There are also multiple reports of a Native American woman with long, straight, glossy black hair being spotted after hours in the man-made Creation and Cosmos cave, which houses the museum's extensive collection of ancient tribal pottery. Some staff members believe she is a benevolent spirit that watches over the crafts made by her people.

I had some downtime during my field trip, so I tracked down the Institute's resident folklorist, Rhett Rushing, and told him about Everyday Paranormal. "We'd really love to do an investigation here," I explained. I figured I would be in for some serious red tape, but Rhett surprised me with his enthusiasm. He had encountered some of the ghosts himself, he explained, and he was eager to see some hard data to back up his personal experiences. Within a week, he had given us the green light to bring a team of investigators to the Institute. It was a warm evening in June when we packed up our equipment and headed out for a night among the artifacts.

By this point, our high-tech arsenal was expanding and we were experimenting with new, more sophisticated equipment. In addition to our digital audio recorders, video equipment, infrared and thermal-imaging cameras, and data loggers, we were also using K-2 meters. These highly sensitive souped-up EMF meters use blinking lights to signal increases in electromagnetic energy, enhancing our ability to do real-time analysis and to home in on active areas.

A few days earlier, we had talked at length with Rhett and his fellow staff members to collect background information about potential hot spots in the Institute. This also helped us to find out what types of paranormal activity we would need to be on the lookout for during our investigation. One of the security guards mentioned that he had seen the apparition of the Native American woman again recently.

"I was on my rounds and I walked into the cave," he told us. "There was this woman in a buckskin dress with fringe lying on the floor surrounded by a bunch of bags. My gut reaction was, 'Oh, no, I just walked in on her.' I said, 'Excuse me!' and turned around. Then I remembered that we were closed and that she shouldn't be in the museum anyway. So I turned back to her and said, 'I'm sorry, ma'am, but you're going to have to leave.' The next second, I was talking to empty space!"

His description sounded uncannily similar to Brad's experience with the woman in the red dress and hat in the Garden District. Unfortunately, both were just personal experiences. We couldn't correlate them to a spike in EMF, a flicker of the lights, or an EVP to prove to skeptics that the apparitions were not triggered by, say, an ocular migraine, a prescription medication, the power of suggestion, or wishful thinking.

But there was another piece of background information that fascinated us and that *did* pass as hard evidence, in our opinion: a digital color photograph that Rhett Rushing had taken about a year earlier. It was of the Institute's famous horse-drawn hearse. At the left-hand side of the photograph, clearly visible, stood the semitransparent figure of a short, stout man in a black suit and white shirt. He wore a handlebar mustache, his black hair slicked back from his forehead, and nothing but socks on his feet. His features were too blurred to make out his expression, but they definitely looked human to us. Rhett assured us that the room had been empty when he had taken the photo. We examined the space to make sure that there was no reflective glass or mirrored surface nearby that might have created the effect accidentally. We didn't think the figure could have been the result of a glitch like an anomalous mist caused by the photographer's breath on a cold night (it wasn't that cold in the Institute and the humidity is carefully regulated to preserve the artifacts) or pixilation (the tendency of digital cameras to break objects into blocky shapes) because everything to the left and the right of the man was in perfect focus. There was no distortion or blurring. Besides, the figure was pretty detailed.

We knew from our research that the dead were sometimes buried without shoes in past centuries. The practice was common because only the upper portion of the casket, showing the deceased's face and torso, would be opened for viewing and because shoes were too expensive for most people to throw away with a corpse. The figure's old-fashioned clothing and stocking feet suggested to us, just as it had to Rhett and his colleagues, that the man might be the ghost of one of the hearse's former passengers.

Based on the photo and the rest of the input we got from the staff, we decided to break our team into three smaller groups that would rotate through three primary areas where paranormal activity seemed to be concentrated. We started with a nineteenth-century sharecropper's cabin that had been moved from eastern Texas completely intact to the Institute's massive interior. It had a small covered front porch held up by the original wooden support posts, where employees and visitors claimed to see the ghost of an elderly African-American man.

Obviously we can't get inside a ghost's head or see things from its point of view, but we do our best to read up on relevant background information before every investigation. If we think we'll be dealing with ghosts from the Revolutionary War period or Colonial times, for example, we'll do a little research on the era, the region, family life, work life, politics, and so on. That gives us a better sense of the life any ghosts we encounter might have led. It helps us get in the right mind-set and know what sorts of questions to ask in EVP sessions. In this instance, we found out that sharecropping was a hard life. Like many others, the Institute's cabin dated from the era following the Civil War, when many freed slaves were too poor to buy property. Instead, they farmed land that belonged to somebody else who had more money, doing all the plowing, planting, and harvesting of cotton, tobacco, or other crops in exchange for keeping a share of the profits made by selling the crop. Sharecroppers still had to take orders from rich, primarily white landowners, and they often made so little money that they had no choice but to borrow from their landlord, which tended to keep them permanently in debt. Many of them spent their entire lives overworked, impoverished, and powerless.

After setting up our equipment inside the cabin, we launched into our first EVP session. "Is anyone here with us?" Brad asked. "Did you own this cabin? What is your name? What crop did you grow? What year do you think it is? Did your family live here with you?"

Nothing happened. Not a shred of activity, not the mildest spike in temperature, not the briefest flash on the K-2 Meter. We were off to a disappointing start, a fact that the recordings we reviewed later bore out.

Of course, the fact that we failed to gather any data is far from definitive proof that an area's claims of paranormal activity are false. Maybe the entity that was haunting the shareholder's cabin simply didn't respond to us. That could be because we didn't bring enough energy or the right kind of energy into the cabin that night. If we don't get a response through polite questioning, provoking sometimes riles a ghost up enough to interact with us, but we haven't collected enough data to know whether *every* ghost can be provoked into activity. Maybe a ghost was there, but it didn't bother to answer us because we asked the wrong questions or because we didn't pique its interest. From what we've experienced, ghosts can be as capricious as living humans.

Fortunately, our next two targets yielded much better results.

The first of them was a room in the Institute devoted to the role of French immigrants in the state's history. After Texas won its independence from Mexico and opened its doors for colonization, a man named Henri Castro seized the opportunity. He led two thousand French-Alsatians to a stretch of land west of what is now San Antonio and founded a town that he named Castroville in honor of himself and which, even today, is dubbed the Little Alsace of Texas in honor of its original settlers. This particular room in the Institute was filled with information and artifacts from Castroville's early days, and taking center stage was the glass-paneled, turn-of-the-century horse-drawn hearse with the mysteriously self-opening doors. It was cordoned off with velvet rope. Despite its beautifully carved and painted wood, its graceful spindly wheels, and its ornate etched-glass windows, there was something ominous and foreboding about the vehicle. It looked like it was waiting—as if at any moment black phantom horses might materialize, their reins held by a black-clad driver, all patiently waiting to carry their unfortunate cargo off for a final somber ride through the city streets to the graveyard.

We examined the coach's back doors to make sure they didn't have a faulty locking mechanism that might cause them to detach and swing open on their own. They didn't. Nor was the coach on an angle that would make the doors fall open if someone left them unlatched. Next, we decided to open the hearse's back doors again just long enough to slip a digital audio recorder inside, which we would leave running throughout the night's investigation. While we did this, we joked about drawing straws to see who should spend the night lying in the back with the recorder, where coffins once rested.

"You know, back in the old days, doctors couldn't always tell the difference between being unconscious and being dead," Brad said. "You really could have woken up in this thing as it rolled through town. And people actually did get buried alive pretty often."

He wasn't just trying to scare everyone. It was true. The practice was so common in the nineteenth century that it inspired horror master Edgar Allan Poe to write a short story called "The Premature Burial" about a fictional example of the phenomenon. There was even a Society for the Prevention of People Being Buried Alive established in Victorian England. People were so terrified of the possibility in the 1800s that coffins were sometimes fitted with bells and breathing tubes just in case.

The recorder was running all this time, though we didn't think much about it. We weren't conducting a formal session, but back at home the next day we replayed the recordings and realized we had caught an eerie EVP that seemed to support exactly what Brad was saying. Just as we opened the doors of the coach and slipped the recorder inside, a voice—hoarse but loud and clear enough to hear over our own chatter—whispered, "I'm not dead."

We recorded an equally intriguing EVP in this area during the formal session that followed. The teenaged son of one of the Institute employees was with us and, as we introduced and numbered our EVP session, he made an offhand comment that caught everyone's attention. "We should be asking these questions in French," he said.

"I speak French," said Hector's wife, Gretchen, who sometimes came along with us on investigations in those days.

"Excellent," we told her. "Why don't you ask a few questions in French?"

"Est-çe qu'il y a quelqu'un ici avec nous?" Is there anyone here with us?

"Je voudrais parler avec l'homme qui j'ai vue dans la photo." I want to speak to the man in the photo.

To our amazement, in reviewing the evidence the next day we caught a male voice that seemed to be speaking another language. We could hear him talking, but we couldn't understand what he was saying. We got Gretchen into the lab pronto and she translated for us: *"J'ai appris frère est ici."* I understand a brother is here.

Could it be the disembodied voice of a nineteenth-century French-Alsatian resident of Castroville? Had he realized that Brad and I were brothers? *Frère*—literally, brother in French—might even refer to a brother in the religious sense. The French immigrants were largely Catholic, and French missionaries were instrumental in spreading Catholicism in Texas in the 1800s. Maybe this was a scrap of conversation from a residual haunting related to a long-ago funeral, a statement from a family member saying he understood that an authorized member of a church was around to perform funeral services or even that the deceased's brother had arrived. Then again, we knew that when settlers move to a new area and take their native tongue with them, the language develops idiosyncrasies in pronunciation, accent, and slang. Words and expressions are sometimes borrowed from other languages spoken nearby. Maybe Gretchen had missed some of the subtleties of the phrase we heard on the EVP. But even if that were the case, she had no doubt that she was hearing French.

And that was what interested us most. After that night, we realized it would be naïve to assume that every ghost in every setting would be able to understand and respond to questions asked in English. It seemed more logical that spirits would communicate in the languages they had spoken in life. That would be in keeping with the fact that ghosts seemed to retain other physical and emotional attributes from the lives they led. The phantom bartender in the photo from the Harlequin, for example, appeared to be wearing his black jacket and bow tie and to have his serving towel draped over his arm. The voices from Alcatraz made references to prison life. EVPs we have caught more recently from soldiers refer to battles.

It was a good theory to bear in mind for future investigations. If our research suggested a spirit might speak another language, we needed to do our best to have an investigator on hand who spoke it, too. (We've managed to get some good data to back this up by having Hector ask questions in Spanish during subsequent investigations and catching some great Spanish-language EVPs as a result.) We also realized that during EVP sessions we needed to avoid current slang that might not make sense to a person who lived during the 1700s or 1800s and to focus on using words and phrases that a ghost from another era would be able to comprehend easily.

Our third target was Creation and Cosmos. Designed to mimic a real cave, it led visitors through dark, narrow, winding halls lined with glass cases of pottery from the Caddo and other Mezo-American tribes dating back to the eleventh century. Just like the antique hearse, the cave looked like an inviting setting for ghosts—or at least for ghost stories. It was possible that people just liked the spooky atmosphere here and let their imaginations get the better of them.

Then again, Native American tradition is rooted in a powerful belief in spirits and the Caddo tribe is no exception. The Caddo people figure prominently into the Texas heritage the Institute showcases. You might say they were the state's very first residents: Like several other tribes, they are descended from the early peoples who lived on the land that is now Arkansas, Louisiana, and Texas as far back as 800 BC. Their ancestors befriended European explorers as early as 1540 and stayed on good terms with them, despite the havoc white visitors sometimes wrought in the form of disease and warfare.

We knew that the Caddo Nation, now centered in Oklahoma, had sent elders to bless this pottery and appointed a spirit to protect it when the exhibit first opened. That act had given rise to the staff members' conviction that the mysterious dark-haired woman was a sort of ghost guardian for the collection, manifesting now and then in human form.

We conducted our first EVP session in the cave speaking politely. "Tell me what your name is. Tell me how long you've been here. Did somebody put you here to protect this pottery?" We followed this by experimenting with a second session that included moderate provoking. Brad threatened to break the glass cases and take the pottery home to sell or to use himself.

Both approaches worked. Back at home, we uploaded the audio files to our computers and found a Class A EVP of a female voice saying what sounded like *"Ku-ah-at"* early in the first session, while we were discussing the difference between a shaman and a medicine man. We did some research online and in the local library and discovered to our amazement that the phrase meant "Welcome" in the Caddo dialect. It reinforced our new line of thinking about spirits using their native languages and it fit neatly with the idea of a friendly protective spirit presiding over the artworks and welcoming in those who wanted to admire them.

Later on in this same session, we were talking about cowboys and Indians, and Brad was goofing around, singing the old 1980s pop song "I Wanna Be a Cowboy." We all chimed in on the chorus. Though we didn't hear anything at the time, when we played the session back, we had one of those "Holy shit! What is that?!" moments. Over the sounds of our own voices, we heard a soft, steady drumbeat and a singsong chant, as if someone we couldn't see were accompanying us.

Finally, when we ended our second session we walked out of the cave as a group and left our recorders behind, still running. In reviewing the evidence, we heard ourselves heading out of the area and then less than a minute later—when we must have been about two hundred feet away, our voices dying out—what sounded like a wooden flute playing a slow, wistful tune reminiscent of Native American music.

We did a search online and found a Caddo song with almost exactly the same tune called "The Bear Ghost Dance." The lyrics discussed spirits and the accompanying dance was performed traditionally to summon long-dead ancestors to a party the tribe was hosting. We found one more interesting fact: Caddo legend holds that the tribe first appeared on the earth by emerging from a cave and that one of the items they carried with them was a drum.

For us, the EVPs from the cave were exciting finds on two fronts. First, they were the best nonvocal EVPs we had caught to date. Second, they seemed to link the ancient to the state-of-the-art, connecting thousands of years of Native American folklore with evidence captured using modern technology.

Brad and I were convinced that the artifacts in the museum rather than the space itself were inspiring the hauntings. We call the notion of things (as opposed to places) being haunted Attachment Theory—the concept that an animate object which was important to a person in life can retain that person's energy after his death as a sort of residual spirit presence. This idea may explain why ghost sightings are common in flea markets, antique stores, and museums. In some cases, collections of "active" objects might create a pool of enough residual energy to spark significant paranormal activity.

We headed back to the Institute a few days later for "the reveal"—the evidence-reviewing session with the staff and got an overwhelmingly enthusiastic response. Granted, many of them would have remained convinced that their workplace was haunted no matter what we had found. The best thing to grow out of our investigation was a strong-lasting professional relationship with Rhett Rushing and his colleagues. In fact, Rhett invited us to return later that month to speak at the Institute's annual Texan Folklife Festival. We weren't sure how people would react to us, but Everyday Paranormal drew a huge turnout, including members of the local media. They followed along as we led ghost tours through the areas we had investigated and played the evidence we had collected for our audience.

The festival, which draws crowds of half a million people, was a great experience for us. We met enthusiastic aspiring ghost hunters, who signed up to join us on investigations of famously haunted places like the Myrtles Plantation and Victoria's Black Swan Inn, which you'll hear more about later. We got to answer questions, heard exciting new stories of paranormal activity, and found out about Texas landmarks we needed to investigate. It reminded us that ghosts have played a unique and important role in the history and folklore of Texas—and that they are still a vital part of our culture, as fascinating to people today as they were hundreds of years ago.

14
Library Specters

BRAD: The Dienger Building in Boerne, Texas, has been many things in its long and illustrious history: a general store, a feed store, a bank, a restaurant, a real estate office, a grocery, and, since 1991, a public library. Built in the 1880s by German immigrants Joe and Ida Dienger, who raised seven children in it, the two-story limestone structure with Victorian gingerbread details looks every inch the classic haunted house it is rumored to be.

Even as kids, we heard the place had ghosts. When we were in high school, playing football against Boerne, we would drive past the Dienger Building on South Main as we headed to and from games, and wonder if the stories were true. Naturally when we formed Everyday Paranormal, we called the library and offered to investigate for them. Though the building is mentioned in countless books about haunted places in Texas, the staff informed us that we were the first paranormal team that had offered to check out the many claims of activity in the library.

The staff members liked the idea of ghosts haunting their workplace; it made the library that much more interesting. They also liked the idea of having us investigate to see whether our evidence would validate the claims, and they gave us an open invitation to bring in our team and equipment any night after the patrons had left for the evening.

The Dienger Building, an official Texas Historic Landmark recognized in the National Registry of Historic Places, sits in the heart of Boerne's quaint, picturesque town square, tucked in among antique shops and atmospheric inns. We arrived after nightfall with Hector, Mike Berger, two Everyday Paranormal newcomers named Brent and Lou*, and K-9 team member Kito, the same five-year-old Boxer who had accompanied us to Laurel's house at the edge of Enchanted Rock. We began the night's work by speaking at length with Adult Services librarian Natalie Morgan, who told us that she and her coworkers had all run across what they thought was ghostly activity at one time or another. They would often walk through the rows of books and discover that a volume had fallen off a shelf, though nobody had been in the room.

"Is it possible that the books were put back on their shelves carelessly?" we asked. Maybe they were perched near the edge, about to fall anyway with a little help from gravity.

"Definitely not," Natalie told us. The librarians made a habit of patrolling the rows to make sure all the books were neatly aligned and pushed well back from the edges of the shelves. "That might be the case if it had happened once," Natalie said. "But it happens pretty regularly."

What's more, from time to time Natalie felt sure that somebody was standing near her even though she was working alone. When she first joined the library staff, she assumed she was the only one who felt it, but gradually by she heard more and more colleagues describe the same sensation. It happened most often on the back staircase (staff members had seen shadows flitting through the dark, heard disembodied voices, and occasionally glimpsed full-bodied apparitions here) and in the basement. It's unusual for a nineteenth-century Texan building even to have a basement. Most were built without them—and many buildings still are—because the soil tends to be either so hard and rocky that it's nearly impossible to carve out or full of shifty, unstable clay. If not that, then there's limestone under the construction site, which would tend to hold any water that leaked in, just like a swimming pool would.

The library building was unique in another respect, too: It stayed in the Dienger family until 1967, when a group of nine local businessmen purchased it and found a cache of some two thousand deer and moose antlers in the basement—trophies from Joe Dienger's many successful hunts. That same year they opened a pioneer-style restaurant serving Hill Country cuisine and called it The Antler's Restaurant in honor of the find.

We ran our first EVP session in an upstairs meeting room that had once been part of the Dienger family's dry-goods store and was now said to be a hot spot for paranormal activity in the building. Joe and Ida Dienger were strict teetotalers and, after The Antler's Restaurant's management team opened a bar in this room, agitated, poltergeist-like activity started occurring. Glasses and bottles would inexplicably fly off shelves and tables and crash to the floor, according to patrons and members of the waitstaff who had witnessed it. Some of the employees reported seeing a male apparition walking back and forth in front of the bar. It eventually came to light that the room had once been the Diengers' master bedroom and that Joe had taken his last breath here. The staff wondered, as did we, if the paranormal activity suggested that Joe was unhappy about people serving and drinking alcohol in his bedroom and was venting his displeasure even after death.

We decided to try some provoking in the hopes of goading Joe or Ida Dienger into responding, presuming that they were the ghosts in residence, as the library staff believed.

"Is this you, Mr. Dienger? Mrs. Dienger? We're all going to have a round of beer. We've got whiskey, too. We'll all get drunk tonight. How about it? Does drinking upset you? I understand that you don't like alcohol. Why is that?"

The ghost or ghosts didn't take the bait here, though. Not a scrap of evidence turned up. It was hardly an auspicious beginning after so many years spent wondering about the famously haunted building.

We had much better luck on the back stairwell. First, Kito made a "hit" on it, becoming visibly nervous at the foot of the stairs. Though he didn't panic as he had done in Laurel's house, he froze at the first step, his ears perked up and his eyes fixed on a point about halfway up, as if he were reacting to something none of us could see. We eventually coaxed him up to the landing, though he hesitated with every step, behavior he didn't exhibit in most other parts of the Dienger Building.

It was the first of three places where Kito made hits—and we collected supporting evidence for all of them. The piece from the stairwell was the most interesting. As we watched the playback of grainy greenish-black infrared footage from the DVR that we had positioned on the stairs, I saw myself standing on the landing saying, "I need you to show yourselves. Make a sound. A knock.

Footsteps. Talk." Suddenly a young boy's high-pitched voice became audible over my own. It blurted out excitedly, "See? See them at the top of the stairs?"

The child seemed to be commenting to someone else like himself about *us,* as if he had caught sight of our team and wasn't sure what to make of us. Could this be data to reinforce our theory about the existence of parallel universes? We wondered, just as we had when we heard the EVP saying "Ask if they're real" at the Freeman Coliseum, if this EVP was the voice of a spirit that existed in another dimension and was temporarily able to catch a glimpse of ours. If so, what opened the window between the two dimensions? Why did it close again so quickly? Our audio recorders didn't catch the voice, which presented another puzzle. The two devices used similar technology, were equally sensitive to sound, and were operating near each other. Why did the video camera pick up the noise while the audio recorders missed it?

The EVP also suggested that at least some of the activity in the library qualified as intelligent haunting. This spirit had the capability to look at us and react. The investigators were the ones to be *investigated,* from its perspective.

Next, we moved to the basement. Kito didn't like it here at all, and I couldn't blame him; it was a creepy setting. I could see why the staff got spooked when they worked down here alone. The space still had its original dirt floors, stone walls, and ceilings so low that my baseball cap brushed the exposed insulation and piping overhead as I walked around. It was orb central, too—dust particles everywhere, kicked up with each step we took. As I mentioned earlier, we've never put any stock in the notion of orbs constituting paranormal evidence. Those opaque little circles that sometimes show up in both digital and traditional film photographs, appearing to float through otherwise ordinary rooms, are just bits of dust or moisture—and as far as we're concerned, they're a nuisance. They get in the way of collecting legitimate evidence on video and camera. Fortunately, they didn't prevent us from collecting a great piece of *audio* evidence that night.

Barry and I were working in neighboring sections of the basement, separated by banks of insulation and able to hear but not see each other, when I heard someone whisper.

"Barry, was that you?"

"No, but I heard it, too," he said.

Our entire team was in the basement at this point, all inspecting various corners of it, and every one of them heard the whisper. None of us could make out the words, though. Later, we played the evidence and found an EVP of what sounded like a child's voice at the exact moment we heard the noise in real time. But the words were high-pitched and unintelligible. Remembering the multilingual evidence we had collected at the Institute of Texan Cultures, it occurred to us that disembodied voices in this particular building might be speaking German because this was a German-speaking settlement a century ago, and the Diengers were no exception. We took the recording to a German teacher at the local high school the following week and let her listen to it. She apologized but told us she couldn't make out the words, either.

Then we tried another tack: We slowed the recording down and eventually we heard the voice say clearly in English, "Bet you can't find her!"

We later learned that the Dienger children had often played

hide-and-seek in the basement. That meant we had an exciting piece of evidence—a VP (a voice audible in real time to the average human ear), supported by an EVP (a recorded paranormal voice audible with the assistance of technology), supported by historical information. It also opened up a new angle to investigate: Could we assume that paranormal speech would mirror the normal speed of a living person's conversation? The speed at which we speak is tied not only to factors like confidence, emotional state, and birthplace (we Texans tend to speak more slowly than New Yorkers, for instance), but also to our need to breathe. Ghosts wouldn't have that need. What were the ramifications of this for those of us collecting EVPs?

At about the same point in the basement sessions, we caught a scrap of conversation between a man and a woman. Try as we might, we couldn't clarify or amplify the sound enough to understand the words—only the distinct pitches of a male voice making a comment and a female voice answering.

Next, we moved back to the first floor, where Barry examined the hallway that separates the old general store from the main library. We zeroed in on this area for several reasons. Shortly after we had arrived earlier in the evening, Brent had opened the door for a woman in an old-fashioned dress with her hair in a bun, assuming she was one of the librarians accompanying us on our investigation. About an hour later, I had glanced down the hall and thought I caught sight of a figure of a woman in a white dress moving through the darkness and then vanishing. Neither Brent nor I knew about the other's personal experience until we took a break and compared stories later. It sounded as if we had seen the same apparition.

It wasn't long before Barry, too, spied the woman. This time she walked from a door at the left side of the hallway to a bookshelf on the right—and seemed to disappear into it. He yelled to the rest of us and immediately started snapping photos.

"Okay," he announced. "I know you're here. On the count of three, I want you to appear in my picture. Ready? One, two, three." And he started snapping away.

Yeah, right. *That's* gonna work, I thought skeptically.

But it did. We went over the dozens of digital shots later onscreen and, sure enough, one revealed a black figure the size and shape of a human being exactly where Barry had seen the apparition. We're not the types to stare at inkblot tests and say we see elephants in them, but this sure looked like a woman in a dress to us. We thought we could even discern an old-fashioned leather boot at the bottom of the shadow. The figure appeared to be stepping out of a doorway, behind Brent and in front of me. Both Brent and I were in perfect focus. The clarity of the image and the fact that the phantom appeared between the two of us suggested to us that the anomaly was no simple technological glitch. It wasn't, for instance, an amorphous blob accidentally generated by Barry's thumb grazing the frame as the shutter closed. We deemed it solid evidence to support the interesting personal experiences that three Everyday Paranormal crew members had encountered in the area. And it actually appeared on command—a first for us at Everyday Paranormal.

We also caught an image of a strange figure at the end of one of the bookshelves. It was almost white and seemed to have human facial features above what might have been an old-fashioned bib dress. We had never seen anything like it before. Nor have we since. We still aren't sure what to make of it, but it looked like a person.

And there was even more evidence. We ran one of the night's last sessions in what's known as the Genealogy Room and caught a very clear EVP of someone picking out a cheerful tune on a banjo, though there was no music playing in the building at the time. Was it a residual haunting from the days of The Antler's Restaurant? Had one of the many Dienger family members who lived here over the years, or possibly a visitor of theirs, played a banjo? We never managed to find out.

Last but not least, we encountered a version of the phenomenon so familiar to the library staff. We found a book teetering at the edge of a shelf. Remembering Celeste the psychic and the WOAI disaster all too well, we stopped the minute we discovered it and asked everyone on the team to review their digital photos. Fifteen minutes earlier, all the books had been aligned. The photos on everyone's camera confirmed it. We retraced our steps to make sure that none of us had been in the main reading room when it happened. Unfortunately, we hadn't left a video camera trained on that particular shelf, so we couldn't prove the book had moved without human intervention.

We returned to the Boerne Public Library a few weeks later for a second investigation and turned up even more evidence to support our conviction that the beloved landmark was haunted. The most interesting event happened at the end of the investigation. We had brought Kito again and he had made a hit on a bathroom at the back of the building's first floor. We were wrapping the investigation up and, as we carried our equipment out to load it back into our cars, we passed that bathroom and heard a noise inside it. We peered in and saw that the water had turned on by itself. We took pictures and video footage. We inspected the hardware to see whether the handles were loose. They weren't. We asked our team members whether anyone had used this particular bathroom. No one had. No one had been in this part of the building. The next day we asked Natalie and her coworkers whether this happened from time to time. Maybe the old building had plumbing quirks. "No," they all said, shaking their heads. "That's never happened before."

We concluded that the Boerne Public Library was a very, very active place for ghosts. It's worth noting, though, that it will *not* be active for book lovers much longer. The town's first library, the brainchild of a local teacher affectionately nicknamed Aunt Jessie, appeared in 1951. Back then a local community-boosting group called The Grange pooled their carpentry and painting skills to turn a bare room of the local fire station into a four-hundred-book library. It was such a success that it outgrew its space and moved to new digs on East Banco Street and then in 1989 to the famous Dienger Building. As of 2010, the library housed a collection of 48,000 items and had once again outgrown its boundaries. Plans to open in 2011, a brand-new library building, to be called the Patrick Heath Public Library, were underway as this book went to press. A group of history lovers have stepped in and are raising funds to turn the historic Dienger Building into a cultural and heritage center. Hopefully, the ghosts won't mind.

Everyday Nutsacks and Other Disasters

In Search of Binky

BARRY: By the time we wrapped up our second investigation of the Boerne Public Library in late spring 2008, we had earned a local following and gotten a number of mentions in local media sources. That meant we were now getting a lot of calls and e-mails from people who had heard about Everyday Paranormal and wanted us to investigate for them. Roughly 80 percent of the contacts were genuine. The people who called really were experiencing what they thought was paranormal activity and they were either curious or worried enough to want someone like us to come in to confirm their suspicions or quell their fears. The other 20 percent, by contrast, were a mix of thrill-seekers, wannabe "ghost whisperers," compulsive liars, business owners hoping to drum up publicity, and seriously deluded individuals. I had a knack for dealing with the public after all my years in EMS work and teaching, and I could usually weed out the fruitcakes. But every now and then, one would slip through the net.

Meredith* was a prime example. She seemed like a normal middle-aged woman when she first contacted Everyday Paranormal. In fact, we had every reason to believe she was credible. She had found out about us through her husband, Dan*, who worked with an old friend of ours. We had met Dan several times, and he struck us as a good guy.

"There's a child's ghost living in our house with us," Meredith told me. "He's been there ever since we moved in more than a year ago. He's not scary, and we don't want to get rid of him. We just want to find out more about him."

"Are you actually able to see the ghost?" I asked her.

"Definitely," she said.

The scenario of a benevolent child spirit sounded plausible enough and the possibility of an intelligent haunting that manifested regularly in a visible form piqued our interest right away. With luck, we might be able to catch this one on camera or video. Judging from Meredith's account, the ghost interacted regularly with both Dan and her. Maybe it was responding positively to them because they were a childless couple about the right age to be parents. Maybe it was just a communicative spirit that would interact with any living person in the vicinity.

We drove out to Dan and Meredith's before the investigation to check out the house and to interview Meredith in more detail. We asked about the history of the place, but she told us she hadn't been able to find out anything concrete about the past owners or whether a young boy had ever lived or died there. That wasn't unusual. In fact, it was rare to find a client who knew the name and background of the spirit that was haunting them. We did a little research on the house and the upscale neighborhood it was located in, but our efforts proved futile, too. Still, we had high hopes that we would learn more once we dropped our electronic surveillance net—as we now called our process of wiring an investigation site with all our recording equipment—over the residence.

We had fine-tuned the operation in recent months and now when we investigated in a smaller space like a private house we preferred to restrict ourselves to a smaller team. This minimized distractions, commotion, false positives, and people tripping over each other in the dark. Plus, most clients didn't want to see a dozen strangers tromping across their living room carpet, dragging cables and tripods.

It wasn't yet dusk when we pulled into the driveway of Meredith's house. When she answered the door, I couldn't help noticing that she was wearing a long, flowing, white nightgown. It seemed a little odd, but maybe she had unusual taste in clothes. We weren't there to judge her wardrobe—just to find out whether her house was haunted.

She greeted us cheerfully and gave us a second tour of her house that doubled as our walk-through and helped our other two team members to get their bearings. Along the way, we took some baseline readings and set up recording equipment in the areas where Meredith described activity happening most often. Was it my imagination or was she slurring her words a little?

We got back to the living room, which would be our base of operations, and set up a bank of monitors that would allow us to keep tabs on what was happening in the various rooms where we had placed our DVRs. As we were running equipment checks, Meredith excused herself and padded off to the kitchen in her slippers.

We soon realized we had positioned one of the video cameras at a bad angle, so I volunteered to run back upstairs and adjust it.

On my way I passed the kitchen. I glanced in absently and spied Meredith standing at the counter pouring herself a generous helping of something amber from a big square glass bottle.

So she was treating herself to a drink. No big deal. Maybe she was nervous about having strangers in her house. Maybe she was more freaked out about her ongoing contact with a ghost than she was willing to admit to us. Or maybe she was apprehensive about the possibility of our confirming her suspicions that she was indeed living in a haunted house. I decided to give her the benefit of the doubt.

By the time I came back downstairs, Meredith had returned to the living room, glass in hand. She was regaling Brad with stories about the ghost. He shot me a quizzical look, eyebrows raised, and I knew exactly what he was thinking: What have we gotten ourselves into?

Before long, Meredith dispensed with her covert trips to the kitchen and brought the bottle out to the living room coffee table where it would be more accessible. She was now working her way steadily through her third drink, growing more and more animated in her descriptions of the ghost with each sip. She drained the glass, refilled it, sloshing a little, and turned to me, leaning in close as if to confide a secret. "Did I menshen, we named our ghosht, Binky?" she asked, eyes wide. Then she gripped my arm and her tone became suddenly serious. "Dijou know he comesh to work with me? Yesh. He ridesh in the car."

I fought the urge to switch off all our monitors and start packing up. It seemed clear to me that the only "spirits" inspiring Meredith's visions had come out of a bottle. But we were there to assess hard data like audio and video footage; her eccentric behavior was basically irrelevant to our work. It was possible that we would find something stellar in the evidence, so we persevered with the investigation, remaining courteous and trying to ignore Meredith's increasingly flamboyant antics.

At one point I was conducting an EVP session, walking slowly through the hallway when Meredith suddenly swept out of the darkness and raced toward me. She swerved at the last second and headed for a corner near the stairs, yelling shrilly, "Binky, you need to come out now! *Binky!!*"

The next day, our evidence review confirmed our skepticism: We had picked up zero activity. We don't usually go in for a face-to-face reveal if we don't turn up any evidence, so I called Meredith to follow up instead.

"Well, you must have just missed him!" she assured me.

"That's possible," I agreed.

"Do you want to come back and try again? Binky can be shy with strangers."

"We've got a pretty full slate, but we'll let you know," I said.

We never took Meredith up on her offer of a follow-up visit and we never found out whether Dan believed in Binky as fervently as his wife did. He made himself scarce during both our on-site interviews and our investigation. We resisted the urge to make jokes about Binky even among team members because the truth was, we felt sorry for Meredith. Did she drink that much every night? If she did, was the ghost an alcohol-fueled delusion? Did she genuinely believe she saw and heard him after she had downed half a bottle? Had loneliness and boredom prompted her to invent a ghost for company? Not having kids of her own, maybe she had created a phantom child as a substitute. What would prompt a person to invent a ghost and then go to the extent of bringing an entire paranormal team in to try to verify the delusion? That's territory we don't investigate.

The Best Photo We Never Got

BRAD: One of the most unusual claims we ever heard came from a woman named Sharon*. Like Meredith's husband, she was a friend of a team member, who asked us to investigate her house as a personal favor. How could we say no?

Sharon was adamant that not only was her house haunted, but that spirits made her levitate and rotate while she was in bed. Her descriptions sounded a lot like scenes Barry and I remembered from several horror movies we had seen, and warning bells tend to go off for us whenever a client's stories are strikingly similar to books, movies, TV shows, or articles that recently appeared in the newspaper. Still, we try to approach every case impartially and with open minds.

Following standard protocol, we headed out to Sharon's and sat down for an interview with her, followed by a walk-through of her house. That's where we spied red flag number two: books about mysticism, the occult, and witchcraft. In our experience, they often signal that the client's "experiences" are stemming from the power of suggestion or wishful thinking. In Sharon's case, every bookshelf seemed to be loaded with them. Of course, it was entirely possible that she had started buying the books as a response to paranormal experiences, maybe in an attempt to gain a better understanding of what was happening to her. And it was also possible that, like us, she was fascinated by anything related to the paranormal—but that her interest, unlike ours, extended to sorcery, voodoo, and the like.

We didn't ask her about her library. It would have been out of line to demand that she justify her choice of reading material to us. It might have put her on the defensive or made her think we were nosing through her belongings. We just made a mental note about all the volumes on the supernatural.

Again, we were working with a small crew, so we assigned a pair of team members to take the first shift outdoors while we investigated the interior. Later on, we switched posts, with Barry and I manning a DVR in Sharon's yard. It was around midnight and we had an infrared lens trained on Sharon's back porch, an area where she claimed to have heard voices and seen apparitions, when something suddenly happened that chased all doubts about Sharon's sincerity out of my mind. As I peered through the viewfinder, what looked exactly like a head and shoulders appeared above the porch railing as if it was peering out at us. It moved from left to right, then it disappeared.

"Did you see that?!" I whispered to Barry

"Yes!" he hissed back. "Did we get it?!"

"We got it!"

It had to be the best apparition image we had ever captured on video! Clearly defined; obviously human in shape and proportions; not flitting momentarily through the scene so that you might miss it if you blinked, but moving slowly and deliberately and lasting no less than five seconds. We could barely keep from shouting the news out to the other team members inside the house. This was too good to be true. . . .

Then we heard a sound.

"Sshhh! Listen!" I whispered.

Something was scraping along the porch railing.

But *that* noise sounded distinctly human, not paranormal.

Barry and I glanced at one another over the top of the video camera. I knew we were thinking the same thing. Fighting down a rising tide of anger and disappointment, we marched up to the house. There, lying flat on her back on the porch for some inexplicable reason, was our client.

"Oh, hi," she said cheerily. "I was just lying down with my dogs on the dog bed."

Neither one of us replied.

"Hey, listen. I'm glad you're here. I need to ask you something. Do you mind if I step out for a little bit?" she continued. "I've got to meet one of my friends down at the cemetery."

"Sure thing," I told her.

Shortly after Sharon left I radioed the rest of the team. "Let's wrap it up," I said. "We're outta here."

We combed through our recordings the next day expecting to find nothing, but to our surprise we discovered a legitimate Class A EVP of a man screaming. The data suggested that Sharon's house actually *was* experiencing paranormal activity. We still didn't buy her tales about levitation though, and we weren't willing to videotape her as she slept to find out for sure. She had exhibited enough bizarre behavior for us to bow out of the picture. Our instincts told us Sharon had no clue that her house really was haunted; that she had made up all the stories she told us. But our job isn't to discredit clients, so we let it go.

Speaking of discrediting people, had Sharon deliberately

tried to dupe us into thinking we had caught an apparition photo? If we hadn't heard her moving and realized who was out on the porch, we might have presented the video footage to her as proof of paranormal activity. Would she have corrected us or been happy to keep us in the dark, so to speak? We are always upfront with clients about the fact that we *never* manufacture evidence. We would rather find nothing than mistake the normal for the paranormal. But every once in a while you run into someone who wants to play "let's pretend" and refuses to believe that you're not in on the game, too.

It wasn't the first time a night's work had proved fruitless and frustrating. It wouldn't be the last. Searching for ghosts requires a lot of patience. You've got to be patient with clients. You've got to be patient while you sit in the dark for hours conducting and recording sessions, sometimes fighting off the nagging suspicion that nothing is happening and nothing is going to. Then you've got to be patient as you play back hours upon hours of recorded video and audio, hoping to catch evidence that generally lasts no more than a few seconds. If you collect three five-second pieces of evidence during a five- or six-hour session, that's a success.

Why don't ghosts appear or speak to the living for longer stretches at a time? We suspect they are only able to harness enough energy to sustain bursts of activity that last a few seconds. From what we've seen, they can repeat the action, but they need breaks in between activity, perhaps to refuel. Maybe someday we will be able to measure and quantify the precise energy levels needed for various types of paranormal activity, and use that information to devise methods of providing the right amount to allow them to manifest for longer periods of time.

"Everyday Nutsacks"

BARRY: It didn't take us long to realize that Lou* had personal problems. He handled himself well during his first few investigations, the Boerne Public Library included, but a few weeks after he signed on to help us find new places to investigate and coordinate with clients, we started to have doubts. He lived about an hour outside of San Antonio, so most of our work with him was conducted over the phone. But even so, it was clear that his moods swung wildly from manic enthusiasm to nonproductive funks, where he would sink into bouts of depression and self-loathing. He hated his life. He hated himself. For some reason, we liked the guy despite all that. We could deal with sluggishness and low self-esteem; it was better than excessive drama or ego. And we felt bad for him, just as we had felt bad for Meredith, so we gave him pep talks and assured him that he was a valuable member of the team.

When Lou finally called to bow out of Everyday Paranormal, I wasn't surprised. "I'm really sorry, guys," he said. "I've just got too many other commitments. Too much going on. But, hey, I wish you the best."

"You, too," I said—and I meant it. Neither Brad nor I thought much about blue Lou until one afternoon when we were going over the investigation videos we had posted on YouTube. There, among our own segments, was a clip we had never seen, titled "Everyday Nutsacks."

"What's that?" I asked, clicking on it.

A guy in a rubber Halloween mask was gyrating and gesticulating wildly, hamming it up for the camera. He wore an official black Everyday Paranormal T-shirt with a pillow stuffed inside it.

"Hi, y'all!" he drawled in an absurdly exaggerated Texas accent. "Ah'm Nad Nutsack. Ah like tuh hunt ghosts . . ."

"What the hell?" I burst out.

"I know that voice!" Brad exclaimed. "That's Lou!"

The man walked off camera momentarily then returned decked out in camouflage from head to foot, wearing another leering rubber mask.

"And Ah'm Larry Nutsack! Ah like tuh hunt an' fish an' . . ."

Brad and I watched the video, speechless. So this was Lou's time-consuming commitment? The reason he was too busy to work with Everyday Paranormal anymore? We wondered what had we done to inspire a parody like this. Weren't we the ones who had forgiven Lou again and again for his pessimistic attitude and sloppy work? Hadn't we tried to help him out, even when he didn't deserve it?

I had kept our old buddy's phone number on hand, so I made a quick call to him. "Listen, we saw your little video. It's pretty funny . . ." Then I gave him an ultimatum: "Take it down or Brad and I will drive out to your house and personally make sure you do."

He was all apologies. "It was just a joke," he backpedaled. "I was just goofing around. Just havin' some fun. I'm sorry. Don't worry. It's gone!" He followed that up with a sob story about how hard things had been for him since he had left Everyday Paranormal and how much he missed working with us. Brad and I actually felt bad enough to invite him to come back as a team member. Keep your friends close and your enemies closer, as the old mafia saying goes. Maybe we would be able to keep an eye on him if he were inside our ranks again.

So "Blue Lou" rejoined our team and started accompanying us on investigations once again. For a while, he handled himself well just like he had at the outset. But before long, he was back to his old tricks. We would give him an assignment, and it would never get done. Clients' calls didn't get returned. Basic logistics for investigations didn't get coordinated. And finally we got a déjà vu–inspiring call: "I'm really sorry, guys. I've just got too much going on."

We parted ways for the second time, and once again Lou decided to trash us at every opportunity. He would log on to paranormal chat rooms and Web sites using thinly disguised pseudonyms and make veiled references to the fact that he had worked for us and knew the "real dirt" about our methods. We shot back responses warning him that his aliases weren't fooling anybody. He started his own competing paranormal venture, but soon enough he lost interest and drifted away.

We never figured out the root of Lou's backstabbing and bitterness, aside from the fact that he tended to be cynical about everything. We knew we would never have universal popularity. Who does? Still, the Everyday Nutsack incident was a nettling reminder of how much vindictiveness there is in our field—and how counterproductive it can be. It made us resolve to put competitiveness aside and to focus on cooperation. After all, that's the best way for all of us who care about paranormal research to help advance the field.

16

Sounds from Beyond

BRAD: The weathered two-story farmhouse in Victoria, Texas, had stood empty for decades, its old wooden floorboards and railings deteriorating as the years passed. The owners, Jack and Bonnie Anderson*, were distant relatives of Barry's. Bonnie had actually grown up in the cavernous barnlike structure in its better days and felt a sentimental attachment to it. So, though the building was no longer habitable, rather than demolish it, she and her husband moved the entire house to another location on their property near the Texas coast and built a more modern home on the foundation of the old one.

As far as Bonnie knew, no one in her family had ever encountered a ghost in the farmhouse when she was a child living there. But ever since she and Jack had moved into the newly constructed replacement house, a number of inexplicable incidents had taken place. They would sometimes leave small household objects lying on a counter or table, and return after the house had been empty for several hours to discover that the items had been moved. At other times, they were sure they heard people talking, even though nobody was around but the two of them.

They asked for our opinion and we told them we thought the

activity they described might be an example of Renovation Theory in action, just as we had suspected at the house in New Orleans' Garden District. It was possible that by lifting the old dwelling off its base and relocating it, they had inadvertently disturbed spirits that had been lying dormant there. They might simultaneously have tapped into a source of energy in the ground that was now providing a battery of sorts for paranormal activity. Maybe there were high radon levels in the soil; an underground stream might flow beneath the foundation. If we had decided to feature the property in *Ghost Lab*, we would have brought in experts in ground-penetrating radar technology and contacted geologists for background on the soil composition of the area to see if we could pinpoint an energy source. But we didn't have access to resources like that when we found out about the Andersons' farmhouse, so we had no way to find out what role the land under the farmhouse might be playing in the haunting.

The property and the historic farmhouse had been in Bonnie's family for generations and she had a hunch that the ghosts might be her own ancestors'. She was hopeful that we would be able to turn up evidence not only to confirm her belief that the place was indeed haunted but which might also help her to identify the specific ghosts in residence.

We brought Jason, Hector, and Hector's wife, Gretchen, along when we investigated the Andersons' property. We asked if we could check out the old farmhouse first.

"Sure," Bonnie told us. "Just watch your step."

We realized what she meant the second we got inside. The deserted house was a fascinating but treacherous place. The stairs were crooked and angled, and when we reached the top, we had to take care not to fall through the gaping holes in the battered floorboards. We walked through it gingerly to get our bearings, took some baseline readings and set up monitoring equipment, then we split into two teams—Gretchen and me as Team A, and Barry with Hector and Jason as Team B.

We had recently encountered a ghost that haunted the grounds surrounding a farmhouse—running invisible through the dry leaves and slamming the old-fashioned garage door—so we now made a point to include the exterior of houses in our investigations. Barry, Hector, and Jason started their work outdoors, while Gretchen and I headed to the second floor.

I switched on my recorder and began my first EVP session, stating my name, the date, and the location. "Is there anybody here?" I asked aloud to one of the empty bedrooms. "What's your name? Did you live here? When? Are you a member of the Anderson family?"

Gretchen and I had only been in the room a few minutes when all of a sudden from downstairs, we heard the thud of heavy boots walking across the floor. *Boom, boom, boom.*

"Can you hear that?!" I asked.

Gretchen nodded, wide-eyed.

"Barry!" I called over the walkie-talkie. "Are you or Hector walking around on the first floor?"

"Negative," he responded. "We're still outside."

The footsteps reached the bottom of the staircase and started climbing steadily up the steps toward us. Then they stopped at the top of the stairs. We waited, fascinated, but no one appeared. We peered out into the hallway, but there was no one there. We checked our equipment and sure enough, we had caught the noise on both

DVR and digital audio recorder—excellent correlative data to support what we had both just heard in real time.

Next, we moved on to what had once been a baby's room. A rusted antique iron crib used by Bonnie's ancestors (she couldn't remember whose it had been) stood against one wall. Again I turned on the recorder. I stated that this was EVP Session Number 2, Anderson farmhouse, Victoria, Texas.

"Are you here? How old are you? Did you grow up here in this house?"

To my amazement, the crib started to tremble, at first subtly then with increasing intensity until it was shaking violently all by itself.

"Brad!" Gretchen hissed. "Knock it off! Why are you doing that? Are you trying to freak me out?"

"I'm not doing anything!" I told her, staring at the crib in amazement. "I'm not even moving. My arms would have to be eight feet long to make that crib shake!"

The frenetic action lasted between ten and fifteen seconds, then stopped as abruptly as it had started. We checked our DVR and there it was: proof positive that Gretchen and I hadn't imagined the unusual activity.

It looked like we were encountering a poltergeist. Poltergeists (from the German *poltern*, which means noisy or rumbling, and *geist*, which means ghost) are spirits that have a unique ability to interact with the material world. They can make loud noises, cause objects like furniture to move and doors to slam or fly open, and sometimes push or pull people. When you combined the two pieces of evidence we caught with the mysteriously moved household objects Jack and Bonnie had noticed, it suggested that the Andersons might be dealing with a poltergeist that had the unusual ability of manifesting in several places—both the old house and the new one.

Lots of people recognize the term *poltergeist* because it was the title of a popular 1982 horror movie, but accounts of poltergeists have popped up around the world for centuries. For example, when Protestant reformer Martin Luther was living at Wartburg Castle in Eisenach, Germany, he was supposedly tormented by one that threw nuts at the ceiling and made noises that sounded like heavy barrels crashing down the stairs. Luther got so frustrated that he finally heaved his inkwell at the spirit he thought might be the devil, and the stain stayed on the walls for centuries.

Arguably the most famous American haunting involving a poltergeist was the Bell Witch case, which occurred between 1817 and 1821 in Adams, Tennessee. The ghost was said to snatch sugar from the sugar bowls, spill milk, yank quilts off beds, pinch children, and pull hair in the household of John Bell. Bell was convinced that the culprit was the spirit of his spiteful neighbor Kate Batts, who claimed Bell had cheated her in a land deal and vowed on her deathbed to haunt him. General Andrew Jackson was stationed nearby with his troops and, when he heard the story, he decided to check it out for himself. As the story goes, Jackson's horses panicked the moment they reached the property line. It took half an hour to calm them down enough to coax them onto Bell's land.

The tale gets even wilder after that. Supposedly, one of Jackson's men boasted aloud that the Bell Witch would be too afraid to show herself with him around because he was a witch tamer and carried a silver bullet guaranteed to kill evil spirits. Moments later, he started screaming that he was being stuck with pins and, as his flabbergasted compatriots watched, an invisible foot kicked him out the front door. The whole party left at sunrise and Jackson was later quoted as saying, "I'd rather fight the British in New Orleans than have to fight the Bell Witch." We've had a whole lot of experience with ghosts and what we have seen makes activity that extensive hard to believe. But, hey, you never know.

One school of thought holds that poltergeists aren't ghosts at all but a phenomenon called recurrent spontaneous psychokinesis or RSPK, which involves an intensely repressed living person (usually a child or teenager) inadvertently channeling their rage, fear, or resentment into moving objects by mind power or force of will alone. But in a house that had been deserted for decades? It seemed unlikely. Besides, neither Gretchen nor I were deeply stressed or emotionally anguished; we were just keyed up and anxious to find ghosts.

Skeptics say a strong air current, seismic activity like tremors in the ground under a house, or even static electricity can cause frenetic movement of inanimate objects that gets misinterpreted as paranormal. However, in the case of the crib, nothing else in the room was moving. Gretchen and I would have felt the floorboards shaking under our feet if there had been a seismic tremor. We would have felt a gust of wind from an open window. As our DVR footage proved, neither explanation applied to the phenomenon we witnessed.

Some people say poltergeists are the worst kind of ghosts—demonic entities that can go so far as to fill houses with foul smells or to bite and claw humans. But our experience suggests they're more likely just mischievous, playful, or frustrated spirits trying to communicate.

Another ghost-investigating team might have drawn certain conclusions from the crib and the heavy footsteps. They might have surmised that this was the spirit of a traumatized child afraid of a father stomping up the stairs, or maybe of a mother who lost a baby and is still grieving over an empty crib from beyond the grave. But there would have been no scientific basis to ascribe specific motives to the paranormal activity we observed in the old farmhouse. Nor did we manage to collect any EVPs in this particular room that provided clues to the ghost's identity or to the true meaning of the shaking crib. For all we knew, the crib was picked at random because it was a handy, heavy object in a room where energy was readily available in the form of Gretchen, myself, and all of our battery-powered tech equipment. We couldn't assume a shaking crib was necessarily a distress signal from the paranormal world. We still knew nothing about who the ghost was or why it was there.

We got luckier with EVPs elsewhere in the farmhouse—and those *did* give us a clue to the identity of at least one of the ghosts in residence. When we played back the audio recordings we had collected from the front room on the ground floor, we were astonished to hear the strains of an old honky-tonk piano banging out a lively tune. It had the tinny sound of the player pianos you see in saloons in old Westerns. In the strictest terms, this is what we would term an ENP (electronic noise phenomenon); it wasn't a voice at all. The melody was distinctive and nostalgic and it ended with a high, rousing final *plink*.

We played it first for Jack, who immediately raised a cynical eyebrow. "Aw, ya'll put that in there!" he said. "You staged it!"

"No, we didn't," we assured him.

Then we played it for Bonnie, and her reaction was as surprising as the ENP itself. She welled up as she explained to us that there had indeed been a piano in that room for many years. In fact, the same piano now stood in the living room of her new house.

"My grandfather used to play that piano all the time," she said. "He played music just like the song you recorded. And he died when he was sitting there at the piano, playing."

We got two more pieces of evidence that reinforced our theory about a poltergeist: I was filming the broken staircase as Jason made his way cautiously down it. He was wearing jeans, a sweatshirt, and a baseball cap backward on his head. As I watched, the cap suddenly flew off his head as if someone had snuck up behind him and popped the bill to knock it off intentionally. It occurred to us that in the past men took their hats off indoors as a sign of respect. Maybe a ghost was making sure Jason didn't forget to mind his manners.

Not long after that, we left. Jason was the last to walk out the back door and down the exterior steps. Again, I was filming him. For no apparent reason, his feet suddenly flew up in the air behind him so that he launched into a sort of spontaneous swan dive and then landed with a splat, sprawled on his face in the muddy yard. His arms were at his sides the whole time, as if he hadn't bothered to put them out to brace himself.

It was hilarious—like watching a slapstick comedian do a pratfall—and Gretchen, Hector, Barry, and I burst into hysterics. "You okay, man?" I asked, trying to stop laughing long enough to make sure he hadn't seriously injured himself. Barry walked over to give him a hand up.

Jason wasn't amused. In fact, he was angry. "Something pushed me," he said. "I felt like I was paralyzed. I couldn't even move my arms to brace myself. It tackled me."

I wouldn't put it past certain people to invent a story like that. It's more dignified than admitting that you tripped and fell down the stairs. But not Jason. If he said it happened that way, it did. And a push in the back was in keeping with the other evidence of a poltergeist that we had witnessed that night.

Conclusion: The Andersons' property *was* haunted. We couldn't say definitively how many spirits were manifesting there, but we were convinced that we had encountered several. Nor, to Bonnie's disappointment, did we manage to identify all the ghosts, though the music was a clear indication that at least a portion of the paranormal activity was connected to her relatives. The music could have been a residual haunting, as could the footfalls. Or it could have been intelligent. Jason's personal experiences, both caught on DVR, indicated that there was at least one highly interactive intelligent haunting taking place at the old farmhouse. What we found particularly intriguing was the fact that the hauntings encompassed both the old house and the neighboring new one. Did it suggest that these were indeed Anderson ghosts and that they had license to roam the property that had belonged to their ancestors?

As you know, we never try to rid a house of its ghosts. In this case, I'm not sure the Andersons would have wanted us to anyway. Lots of people who own land, houses, or heirlooms that have been in their family for generations feel a connection to the past—as if they are surrounded by the spirits of the ancestors who once tilled the soil, walked through the rooms, rocked in the antique rocker, or dined at the old farmhouse table. Bonnie was just surrounded by them a little more literally than most people.

17
School for Scares

BARRY: Being a teacher in the local public school system and having a wife who taught in the same district meant that word soon got around in the education community about what I did as a hobby. Most people were intrigued when they found out, and either asked me to tell them more about ghost hunting or struck up conversations with me about a paranormal activity they or someone they knew had experienced. If their accounts were accurate, a surprising number of our area schools were haunted. In fact, several of my colleagues told me that the building I was teaching in had its own ghosts. At that point, the building housed administrative offices and classrooms equipped for children with special needs, but it had been an elementary school at one time, and some staff members were convinced that the ghosts of former students were still in attendance—responsible for a number of weird occurrences like flickering lights, mysterious voices chatting in one of the boys' bathrooms when it was empty, and rocking chairs moving when no one was sitting in them.

Heather* had recently started working in the building on weeknights as a counselor, and broached the topic of ghosts with me. A scholarly fifty-something who swore she'd had no paranormal experiences before taking this job, she wasn't the sort of person prone to superstition or hysteria. But she had encountered two things recently that she couldn't explain—or forget easily.

"The first happened around nine o'clock at night, and I was the only one here," she told me. "I'm almost always the last one to leave, so I was going through the building to make sure all the doors were locked, like I usually do. I was walking down one of the hallways and I saw an older woman ahead of me—maybe twenty or thirty feet away—walking, too. I had never seen her before and I couldn't figure out what she was doing here, so I called to her. 'Ma'am, can I help you? Ma'am?' She kept on walking, as if she hadn't heard me. Then she turned into a room. I caught up and walked in behind her. I know it sounds crazy, but all the lights were off in that room and there was no one in there but me!"

Heather was genuinely unnerved by the apparition. "The woman wasn't transparent," she said. "She looked and acted just like a real person . . . until she disappeared." I assured her that it wasn't uncommon to mistake a ghost for a living human being. Brad and I had heard lots of similar descriptions about spirits that seemed to be flesh-and-blood men, women, and children based on their physical appearance, movements, and actions. As you know by now, not every ghost floats down a staircase or drifts out of a wall as a pearly white form you can see through. The form taken seems to depend on the area's EMF level, environmental conditions like humidity, and—to some extent—either the ability or whim of the individual ghost.

I also told Heather that, though I hadn't experienced any

paranormal activity while I was at work, some of my students had exhibited odd reactions to certain parts of the building. For example, one little boy who I taught was autistic and he flat-out refused to use the boys' bathroom, where staff members claimed to hear disembodied voices and feel invisible hands tugging on their sleeves and pant legs. Will* couldn't or wouldn't articulate what he disliked so much about the room, but he would panic every time anybody tried to take him near it. He had no aversion to any other room in the school, from what I could tell. Did Will know the stories about the lavatory? Had somebody fueled his fears of the room by warning him that it was haunted? I couldn't say for sure, but I doubt it. And if not, his aversion to the bathroom was all the more intriguing.

Was it possible that an autistic person might be sensitive to paranormal stimuli that the rest of us would miss? Was Will actually hearing, seeing, feeling, or sensing a presence I couldn't pick up? If his brain and senses were hardwired differently from the average person's, did it stand to reason that they might be better-tuned receptors for paranormal activity? And was it possible that ghosts sensed this and tried to communicate with him because of it? Some people believe in the concept of "indigo children"—kids who are born with especially keen sensory perception, which enables them to see ghosts, to read minds, or simply to intuit other people's feelings much more astutely than the average child. But there was no way to test any theories I might have had on a link between autism and receptivity to paranormal input without exposing Will to a room that upset him, so delving into the area further was out of the question.

Heather's second paranormal encounter had also occurred

during her usual nighttime lockup—and a security camera caught the whole incident.

Heather was walking down one of two hallways that run the length of the building when something caught her eye at a point halfway along, where a door was set back into a shallow alcove. When we reviewed the security camera footage we could see what had drawn her attention. The door was open a crack.

"Hmm, that's weird," she muttered to herself, peering at it curiously. "That's supposed to be closed."

A small wheeled cart of the type teachers use to haul books and audio-visual supplies was standing against the wall nearby. Heather grasped both sides of the cart and steered it into the alcove so that it pushed the door shut. The second the cart touched the door, something or someone on the other side growled loudly and angrily. You could hear it distinctly on the videotape.

Watching the video was more hilarious than scary. The moment she heard the snarl, Heather let go of the cart and staggered backward, arms flailing. She nearly lost her balance, then spun around and started racing down the hall away from the noise as fast as she could. I couldn't blame her. It did sound ominous.

We were eager to explore the unexplained goings-on at work—with an Everyday Paranormal investigation. I talked it over with Brad and we got permission from the superintendent of the district to camp out in the building for a night in late spring 2008. We combed the inside of the room where Heather had heard the growl as well as the surrounding area. We opened and closed the door repeatedly. We experimented with pushing, pulling, and turning the cart to see if we could re-create the sound. But we never found any logical, nonparanormal source that would have generated a growl like the one in that alcove.

Nor did our investigation yield anything to rival the menacing sounds from behind the door. However, when we reviewed our evidence the next day, we got some data that correlated with the many reports of activity in the bathroom that had spooked Will so intensely.

Brad was talking aloud, trying to prompt some activity. "Hey, if you're in here, flush this toilet! Make this faucet turn on! Slam one of these stall doors!"

In one of the clearest EVPs we have ever managed to record on video, the high-pitched voice of what was undoubtedly a little boy said, "I'm not back there!"

We met with some of the staff for a reveal and played the video recordings for them. They were stunned. Even the most skeptical among them said we had convinced them that there just might be some truth to the creepy stories they had heard for so long about their workplace. The superintendent was no exception. When she heard the EVP, her eyes grew wide. "What the heck is that?!" she asked, looking aghast.

The contacts we made through that first school investigation opened the doors for investigations of several other schools nearby. The most fascinating of them was Cibolo's Wiederstein Elementary, where my wife worked. One of the oldest schools in the district, it was the focus of numerous ghost sightings. The custodial staff claimed chairs moved back and forth on their own, disembodied voices echoed down the halls, and lockers slammed when nobody was near them.

Construction on the original building had started back in 1955 and concluded two years later, when the school opened as a junior high. It was renamed O. G. Wiederstein Junior High School in 1961 in honor of Otto G. Wiederstein, who taught and served as a principal in the district for fifty years. In 1969, it became an elementary school, teaching kindergarten through fifth grade. Numerous renovations were made over the years, including the addition of a library in 1974, and a new wing on the back of the school in 1990. But by the time we got the go-ahead to investigate, the old place was about to dismiss its last class, giving way to a larger, more modern replacement campus opening nearby.

Contractors were still putting the finishing touches on the brand-new Wiederstein Elementary, but the walls in the old school had already been stripped of student artwork, the books and furnishings packed up and shipped out by the time we showed up, literally the day after the last students walked out the door. The building would later be rededicated as a learning center, but when we arrived it had the forlorn, vaguely foreboding feeling so common in abandoned buildings.

We were planning a low-key investigation—just Brad, myself, one other team member from Everyday Paranormal, and Jenny*, the school librarian, who happened to be a friend of ours. She had worked in the building for years and often heard odd noises when she stayed after hours, so she was curious to see what we might find in the way of evidence.

"I rested up all day so I could stay up all night for this investigation with y'all!" she told us enthusiastically as we made our way through the halls and classrooms, taking baseline readings and setting up what we call a shotgun sweep, where we spread out our monitoring equipment in a wider, more random area rather than in a straight line to create a linear sweep.

We left recorders running at various points and moved on to the library. We were sitting there talking, telling Jenny about our plans for the upcoming Folklife Festival, when we heard an unexpected noise—the distinctive *click-clack* of high heels walking along the hallway toward us. "Who's that?" Jenny wondered. "No one else is supposed to be in the building." We waited, eyes fixed on the door, as the sound drew closer, assuming a school official had dropped in to check on us and would appear at any moment. Instead, the noise passed and faded away down the hall.

All four of us jumped up and hurried over to peer out through the glass-paneled doors. The corridor was dark and deserted. Brad and I split up, each taking a portion of the hallway. Next, we checked the empty classrooms nearby. There was no one out there.

A little while later, Brad was investigating a hallway in the original school wing and decided to try some serious provoking to see if he could get a rise out of anything. It was lined with old lockers and he started pounding on them as hard as he could with his fist. *Bam! Bam! Bam!* The sound reverberated off the walls.

"You get out here!" Brad yelled.

And to our astonishment, something growled at him. We all heard it clearly.

"Oh shit!" Brad said.

"Whoa, look at the time!" Jenny exclaimed. "I really need to get home and check on my kids." Gone were her plans to stay up and burn the midnight oil with us on her first ghost hunt. She simply handed over the keys to the building and told us, "Y'all just lock up when you're done."

Were the two growls related? Were they necessarily the sign of a sinister or even demonic presence in the two schools? We can only guess, bearing in mind that interpreting paranormal actions through the lens of normal human ones might mislead us. From our experience, a shove in the small of the back isn't necessarily a hostile gesture. Nor are children's giggles necessarily a benign, friendly message from the spirit world.

We conducted sessions in numerous areas inside the school and collected enough evidence to suggest that the place was indeed experiencing paranormal activity. I left my audio recorder in one of the hallways and it picked up a booming EVP, as if someone were banging on lockers. We doublechecked to make sure it hadn't picked up the noise during Brad's bout of pounding elsewhere in the building, but the two sounds occurred hours apart.

Even eerier was the singsong chant we caught. It sounded just like a little girl reciting, "One plus one is two plus two is . . ." We were three halls away, more than fifty yards from the recorder, when it picked up the noise. And the words were clear as day, as if a little kid's ghost was just wandering through the halls singing. EVPs of children's voices have always seemed the creepiest to me, and this one sounded as if it had been lifted wholesale from a horror movie soundtrack. I still get chills when I replay it.

We found out later that a young girl had died of an illness decades earlier while she was a student at Wiederstein—not surprising, given the many years the school was in operation and the countless pupils it served. Though she hadn't died in the building itself, a tree had been planted on the school grounds in her honor and still grew there. Was the ghost linked to this tragedy? Was it an unrelated spirit? Could we assume that it was definitely the ghost of a little girl, based on the tenor of the voice we caught? Maybe schools, like other buildings that witness intense emotion and strong memories, tend to become magnets for paranormal activity when the energy levels are right. Of course, there's also the possibility that children are sensitive to ghosts or that certain ghosts are drawn to their energy, exuberance, and open-mindedness. One thing's for sure, back when Brad and I were kids, you can bet we would have loved the idea of attending a haunted elementary school.

18
The Battle over *Ghost Lab*

BRAD: By the summer of 2008, business was booming. We had built up a devoted local following, which meant our days of cold calling were over. People were now contacting us to ask if we had time to investigate the paranormal activity they were experiencing. Barry and I were speaking at more and more local events, and Everyday Paranormal had generated enough buzz that we could pack the house. We still weren't making money from our investigations, but we were doing what we loved and we were getting better and better at it, developing new theories and refining our techniques to test them in the field. Ace Productions* had filmed several of our local investigations as well as an entire investigation at the Myrtles, and someday soon they would have our finished DVD ready. We would be able to promote our work more effectively—maybe even take Everyday Paranormal to the next level. In fact, I was about to call D. B.* for a status report on the DVD when I got an unexpected call from him.

"Uh . . . listen, Brad . . . we need to talk." He stammered and stuttered, then he lowered the boom. "Ace can't invest any more time in you," he told me.

I was stunned. "Why not?" I asked. "Is there a problem with

the footage?" The guy had filmed hours upon hours of us at work, but maybe he was a stickler for perfection and what he saw wasn't passing muster.

"No," he said. "It's not that."

"What then?"

"Well . . . we're not making any money on your project."

"*You're* not making any money?" I asked, dumbfounded. "*Nobody's* making any money from Everyday Paranormal. Not me, not Barry, not the team. This was never about money."

"Yeah, but it's a lot of work," he griped. "We don't want to invest any more time in you guys if we can't get paid for it."

It felt like extortion. D. B. had been full of enthusiasm when he had contacted us initially, full of talk about our potential and how he would help us parlay it into something big. We had bought it hook, line, and sinker. We had footed the bill for Ace's rooms at the Myrtles, bought them meals, paid for their gas, even bought batteries for their video cameras. They had nothing out of pocket except their time. Now D.B. was trying to hold us hostage: No *dinero,* no DVD.

So this was why D. B. had been dragging his feet, letting weeks and weeks slip by without even showing us a rough cut and coming up with a new excuse every time we spoke.

"Look, it just isn't working out," he said. "We're gonna have to drop you."

I wrestled down the urge to explode. Instead of bawling him out, I said through gritted teeth, "Okay. Fine. Why don't you just give us what you've finalized so far and we'll call it quits?"

He hesitated. "Uh . . . when you say *finalized* . . ."

I couldn't believe it. Based on his reaction, my guess was that the guy had done virtually nothing with all that footage he had been hoarding. After a tense few minutes on the phone, D. B. finally agreed to meet us and hand over our tapes.

Barry and I were so mad we hardly spoke during the drive to the neutral meeting point D.B. had chosen: a parking lot in San Marcos. The whole situation was ridiculous. There was something weirdly covert about pulling into an empty lot and sitting there in the front seat, waiting for another car to show up to hand over a bag of goods. I felt like we had stumbled into a scene from a mob movie.

Finally, D.B. arrived, turned off the engine of his car, then reached into the backseat and hauled out a huge sack full of tapes. Sheepish and hangdog, he barely glanced at us as he lugged the bundle over and thrust it at me.

As soon as Barry and I got home, we rummaged through the bag and found the DVD. Watching it confirmed my suspicions. The thing wasn't even a quarter finished. Ace Productions had hardly done any work on it. Now we were stuck with hours and hours of footage that still needed to be edited. We're not editors. We didn't have the faintest idea what to do with it. Mike Berger eventually came to the rescue and put together a respectable amateur DVD, but in the meantime, Ace Productions gave our old buddy Celeste the psychic a run for her money in terms of making our blood boil.

Still, we couldn't let ourselves dwell on the setback. We had the usual overly full plates—growing families and demanding full-time jobs, plus another big investigation we needed to get organized asap.

Next on the slate for Everyday Paranormal was Victoria's

Black Swan Inn, a famously haunted historic landmark dating from the days of the Alamo. A stately white mansion with a big columned porch and a rolling green lawn, the Black Swan sat on the northeast side of San Antonio on a well-known former battleground: site of the bloody Battle of Salado Creek in 1842, where sixty Mexican soldiers died along with a Texas volunteer. Archaeologists had also found Native American relics on the property and, like numerous other sites we had investigated, the land was thought to be an old tribal burial ground. The ghost stories included apparitions of a man who seemed to hover outside a second-floor window looking in, disembodied children's laughter, and repairmen getting jabbed and prodded by invisible sticks or fingers as they worked under the house. We had no idea whether we would be able to gather evidence to support the claims, but in our experience, a history of war, death, and burial in an area *does* tend to be conducive to paranormal activity.

I was especially eager to check the Black Swan out because my wife had seen a ghost there when she was fourteen. The owner's daughter happened to be a friend of hers and invited her to a slumber party at the inn one night. Jessica woke up early in the morning and realized she was thirsty so, without waking any of the other girls, she tiptoed down to the kitchen to get a drink of water. On the way, she passed a butler in the pantry who was cutting up vegetables. Surprised by his formal appearance and not wanting to disturb him, she walked past quietly, got some water, and went back upstairs to find the other girls awake and chatting.

"Boy, your butler comes in to work early in the morning, doesn't he?" she said.

"What?" asked the friend who was hosting the party.

"Your butler. He's already downstairs working."

The girl gave her a blank stare. "We don't have a butler," she said.

BARRY: A few days after the big videotape exchange in the parking lot, Brad and I were doing some prep work for the Black Swan investigation when the phone rang.

"Everyday Paranormal. This is Barry."

"Hi, Barry," said a voice on the other end. "My name is Al LaGarde. I'm a TV producer and I'm looking for someone to help me get background footage for a paranormal reality show I'm working on."

My fear that it was a prank call evaporated as I listened to Al explain the situation. The show in question had had several sites cancel upcoming investigations and needed some replacement material to fill in the gaps. He was hoping we would be able to help him find a few haunted places to investigate. He also said that if we were willing to share some of our techniques, he might be able to offer us a cameo on an episode of the show.

But we were gun-shy after getting burned by Ace Productions. We had already heard empty promises about TV. Besides, our techniques were proprietary. We had worked hard to develop them. Why give them away? But then Brad and I talked it over and eventually decided to give it a try. Our goal was to advance the field of paranormal investigation, and here was a new forum we could use to do that. Besides, maybe it really would open some doors this time.

We invited Al to join our investigation of the Black Swan in July 2008, and crossed our fingers that it wouldn't be a bust. We warned him that the night would be a little atypical for Everyday

Paranormal because we had planned this particular outing as a combination investigation and educational experience, along the lines of the Myrtles trips. We had promoted it during our seminar at the Texas Folklife Festival and invited almost forty ghost-hunting enthusiasts to join us. Involving the public inevitably introduces an element of unpredictability: Based on what we had seen in the past, we would get a mix of people intensely serious about ghost hunting, people who were just there for kicks, and a few whack jobs. That night, the prize for bizarre behavior went to a thirteen-year-old who attended with her sister and who claimed to have medium abilities. She swore she saw a man in one of the bedrooms and launched into a semi-hysterical episode about it.

To our disappointment, her outburst was the most exciting moment of the investigation. We registered a big temperature fluctuation in the yard and we caught one apparition photo, but we came up empty-handed on EVPs, which was unusual for us. It wasn't enough evidence to draw any firm conclusions. Was the inn more hype than genuine haunting? Or did a combination of too many people and too much commotion make it impossible to gather authentic evidence? Given our theory about ghosts feeding on energy, you might think that the greater the amount of human energy, the higher the odds of a ghost manifesting. But it doesn't seem to work that way. If that were the case, the bulk of the world's ghost sightings would probably occur in crowded stadiums and commuter trains at rush hour. Could it be that too many different energy sources confuse ghosts? Is it possible that ghosts are actually present, but we're just too distracted to see or hear them? We're not sure yet, but we may eventually be able to collect enough data to learn the answer someday.

Despite its shortfalls, the investigation at the Black Swan Inn apparently convinced Al that we knew what we were doing. We got a call from him a few days later. "I really like your style," he told us. "You've got a lot of personality and a fresh approach. It's scientific. It's educational. It's new. I haven't seen it applied to paranormal shows anywhere. Do you think you two would have any interest in doing your own show?"

We almost fell off our chairs. We were always bs-ing about how we would do things if the world were watching us. We had poured countless hours into Everyday Paranormal and lost a lot of sleep over it. It would be nice to be rewarded for it. But we had done all that because we loved investigating and we wanted to stay true to our original mission: everyday guys on everyday ghost hunts. Not because we wanted to see our faces on TV.

"We don't want to sell out," I told Al cautiously.

"That's exactly what would make you guys good on the air," he assured us. "Why don't we give it a try and see what happens?"

We liked Al's no-nonsense, honest professionalism from the start. He was everything Ace Productions wasn't. So we gave him the green light to create a short promotional video—called a "sizzle reel" in TV parlance—to shop to the networks based on the footage from the Black Swan Inn.

After our earlier experience with promotional videos, we were impressed that he actually finished the project. When he called a short time later to tell us that several networks were interested in developing a show with us, we were floored. But now we needed a second sizzle reel—something slicker and harder hitting, with more evidence. "Can you find another haunted landmark in San Antonio where we could film?" Al asked.

We knew just the place: the GDC Building.

Guerra DeBerry Coody (GDC) is a communications and marketing firm based in downtown San Antonio's historic Soledad Building. Constructed as the Savoy Hotel in the 1800s, the building became a furniture store, a brothel, the site of a grisly murder by a gang of Mexican outlaws led by Pancho Villa, and eventually a derelict wreck, home to squatters and stray animals. It fell into deeper and deeper disrepair over nearly seventy-five years before GDC came to the rescue, buying the property and giving it a much-needed overhaul. Today, thoroughly modern advertising offices sit cheek by jowl with portions of the old Savoy, which retain their original nineteenth-century fixtures and signage. When you walk through the door separating the two sections, you feel as if you've suddenly stepped backward through time.

No sooner had GDC's renovations begun than work crews started to notice strange goings-on. They heard voices and saw shapes flitting through the darkness. Some refused to set foot inside the building afterward. We thought this might suggest another case of disrupted ghosts like the ones we had seen in New Orleans and possibly at the Andersons' farmhouse. Remember, Renovation Theory holds that remodeling an old building can change its energy level, creating an inadvertent wake-up call for ghosts.

The staff at GDC had asked us to investigate their building once before because employees, too, were convinced that there was more of the hotel's past hanging around than just the wallpaper. Like the workmen, they saw apparitions. Lightweight office equipment like calculators and adding machine moved or turned on when nobody had touched them. People heard voices and laughter in empty rooms. (Incidentally, we featured the GDC Building in Season One of *Ghost Lab,* and found some extraordinary evidence to support their claims.) For our first investigation of the old Savoy-turned-GDC, we chose Halloween night. A few overly festive staffers tried to turn our ghost hunt into a cocktail hour, but we caught a lot of good evidence anyway—EVPs and personal experiences on every floor, starting with the eerie sub-basement where there's a hulking old furnace worthy of a scene from *Nightmare on Elm Street.* Our favorite recording was of a voice saying, "Who's that? Who's there?" after a bout of loud provoking by Brad and me. Not only did it appear to be a direct response to our voices, but it reinforced the data we were collecting to expand our theory about parallel dimensions. Counting the EVPs from the Freeman Coliseum and the Dienger Building, this was the third piece of evidence we had collected that seemed to suggest it was possible that parallel normal and paranormal dimensions existed simultaneously and experienced brief, sporadic overlaps, which tended to confuse the inhabitants of both. When we presented our evidence to the GDC gang, their jaws hit the floor. "Come back anytime!" they told us.

The building was a perfect setting for the new and improved sizzle reel. It had an interesting history, great architecture, and it showcased our hometown nicely. More important, we knew firsthand that it was haunted.

Al was now working with some associates in Los Angeles on our project, and they agreed to send a cameraman to film us during round two at the GDC Building. Again, we lucked out and caught great evidence—more than enough for a new reel to pitch to the networks. We touched base with Al, who was full of optimism after seeing the footage and told us he hoped to have good news for us soon. But as we should have known from our own investigations, you've gotta wait until all the evidence is in before you draw any conclusions. . . .

BRAD: About the same time all these positive developments were happening, we were starting to encounter some unexpected and unnerving activity that had nothing to do with the paranormal world. And if we had thought hanging out at midnight in a creepy old-folks' home or locking ourselves into solitary confinement cells on Alcatraz was hair-raising, they turned out tame compared to dealing with Hollywood heavy-hitters and New York lawyers. Both suddenly started materializing like apparitions, full of contracts, promises, and threats.

Before we knew it, we were getting calls from people we had never heard of in Los Angeles and Manhattan. They were all TV types—and they were all telling us that they owned the rights to any televised program that resulted from our sizzle reel.

Al reassured us that *he* had signed us and had the exclusive rights to shop the concept of a TV series starring the Klinge Brothers and Everyday Paranormal to the networks.

"That sounds good to us," we told him.

However, before long, the gloves were off and we realized we were caught in the middle of a furious turf war—with Everyday Paranormal as the turf. We didn't even have a show yet, but we had a whole bunch of people we had never seen face-to-face who wanted a piece of us if we ever got one.

"Look, it's no big deal," they assured us.

"This happens all the time."

"Don't worry. We can work through it."

Barry and I didn't know what to do or whom to trust. Everybody kept promising to fly to Texas to meet us, but they were all stalling on the paperwork. Nobody would sign anything. And they were never in their offices when we called.

Next, ultimatums arrived: So-and-so would move forward with us, but only if we cut out his competitors. Mostly, everybody wanted Al out of the picture. We didn't. He had discovered us, and we felt loyal to him.

Then we got the doomsday call. "Listen, we're going to give up on your project. It's too much of a nightmare."

Where had we heard *that* before?

We called Al in desperation, and with his usual grace and professionalism, he agreed to bow out. "Let's keep in touch," we said. "Maybe we can work together again in the future."

However, it soon became clear that the producers who were supposed to be championing our project had a few hidden agendas. Namely, the goal seemed to be to take us out of the game so we wouldn't be able to compete with other shows they were shopping.

We were frustrated and fed up. So we reminded ourselves that we didn't care about all this. We would put it out of our minds and go back to doing what we had always done. TV was too much trouble anyway. But the minute we said that, we got another chance: Our territorial would-be promoters had managed to get their own pet projects green-lighted, which left Al free to step back into the picture.

"I've heard Discovery is looking for a paranormal show," he told us. It was almost too good to hope for. Discovery had always been at the top of our wish list.

There was a caveat, though, Al warned. The show would have to be educational and it would have to be scientific. It would have to be accessible enough that anyone could understand it. No nonsense. No drama. It would have to focus on using scientific instruments, gathering data, and analyzing it objectively.

"Do you think you can do that?" he asked.

"Dude," I told him. "That's exactly what we've been doing since the day we founded Everyday Paranormal."

BRAD AND BARRY: That was two years ago. Today, *Ghost Lab* has been broadcast in seventy countries and has reached millions of viewers around the world. We speak at conferences with hundreds of attendees, many of them avid, enthusiastic amateurs passionate about their quest to find answers to the unknown and unexplained, just like we were not so many years ago. Just like we still are, in many ways.

In writing this book, we've thought a lot about all the times in our lives that we ran up against brick walls in our research—all the times we were told "no."

No, you won't be able to find any nonfiction books in the library about ghosts.

No, you can't try a new technique to draw out an EVP.

No, you'll never be able to make a TV show that turns accepted ideas and theories about the paranormal world upside down.

But the truth is: yes, you can.

If you put your mind to it, you can do anything. Even if it's something as completely crazy sounding as building a career around looking for ghosts.

We're living proof.

Ghost Hunting 101
The Basic Tool Kit

Below are twelve basic tools of the trade that we recommend investing in if you want to try your hand at paranormal investigating. You should be able to fit them all into one lightweight duffel or backpack.

1. Handheld digital audio recorder
2. Handheld digital camera with flash capability
3. Handheld digital video recorder with infrared illuminator
4. Handheld digital indoor/outdoor (noncontact) thermometer with luminous display with ambient temperature capability
5. EMF detector
6. Walkie-talkies (if you plan to have people investigating in different areas simultaneously)
7. Lightweight flashlight
8. Extra batteries for all the equipment above
9. Digital watch with luminous display
10. Measuring tape
11. Notepad and pencils

12. Cell phone (don't forget to charge it beforehand) for emergencies. Make sure it's turned off or in airplane mode so it won't send or receive signals that could generate false positives.

If you can afford a splurge, add the following technology:

- Data logger (this can take the place of a digital thermometer; in addition to temperature, it records dew point, humidity, and other environmental data)
- Portable motion sensor
- Night-vision equipment like infrared goggles
- Parabolic microphone
- Thermal-imaging camera

Glossary

Anomaly: An event that can't be explained by normal science or physical laws.

Apparition: A spirit in recognizably human form.

Attachment Theory: Holds that an object that was of particular significance to an individual during his life can retain the energy of that person after his death in the form of a residual or intelligent spirit presence. Some paranormal theorists go so far as to believe that an object can actually be cursed by an owner who leaves residual negative energy attached to it. Flea markets, antiques stores, and museums can house enough "active" objects to spark significant paranormal activity.

Baseline reading: A controlled reading of environmental factors like temperature and humidity taken when there's no paranormal activity happening. Once these have been recorded, they help an investigator to track any unexpected changes that could signify paranormal activity.

GLOSSARY

Cold spot: An area with a temperature significantly lower than the surrounding ambient temperature. When no logical explanation can be found for the differential such as an air-conditioning vent or open window, paranormal investigators believe a spot may indicate the presence of a ghost. The cold spot is thought to be the result of an energy or spirit drawing power from ambient heat in the air, leaving a cold void.

Data logger: A digital information reader that takes continuous snapshots of environmental conditions, including temperature, dew point, humidity, and EMF on a scheduled interval.

Digital voice recorder (DVR): A recording device, usually handheld and battery-powered, that picks up sound and translates it into a digital format that can be uploaded easily to a computer.

Digital indoor/outdoor thermometer: A thermometer that measures air temperature and displays it in numeric form.

Ectoplasmic fog: A paranormal fog, mist, or mass that can be seen in real time by the naked eye; a manifestation of an entity in its simplest visual form.

EMF (Electromagnetic fields): High EMF can occur naturally in the presence of computer screens, cell phones, or power lines and in homes with faulty wiring. However, sudden EMF spikes without electrical explanation can signal the presence of a paranormal entity.

EMF detector (or EMF meter): A scientific instrument, usually handheld and battery-powered, that measures and records/displays electromagnetic fields (EMF).

Entity: A spiritual being that may have been human once, but is no longer alive.

Era cues: Everyday Paranormal's theory that you can elevate paranormal activity from spirits of a certain era by presenting familiar stimuli.

EVP (Electronic Voice Phenomenon): Voices and other noises captured using digital audio recorders that can detect sounds below the threshold of human perception. In the paranormal world, under the correct conditions, these sounds are considered to be voices from beyond the grave. EVPs generally last just a few seconds. They can convey extreme emotion or be spoken in a monotone. Class A EVPs are clearly audible on playback without digital enhancement. Class B is relatively loud and often audible even without headphones. Class C is extremely soft and the words may be unintelligible. They can also be voices naturally embedded in the environment from a previous time and not related to spirits at all.

Ghost hunter: Someone who investigates reports of ghosts and paranormal activity. Synonym for paranormal investigator.

Harbinger: A paranormal prophetic sign foretelling a significant event to come, either good or bad.

Hypnagogic hallucination: Also known as a waking dream, this episode occurs on the threshold of sleep and waking and can be accompanied by temporarily full-body paralysis.

Imprinting (or place memory): The notion that strong memories and extreme emotions can leave a lasting mark on a place in the form of residual energy.

Intelligent haunting: A haunting by a spirit that interacts or responds to living humans in the physical world.

K-2 Meter: An EMF meter that uses flashing lights to signal an increase in electromagnetic energy.

Linear sweep: Involves placing data loggers in a straight line down a hall or stairwell. The data loggers take a reading every few seconds, so if any unknown entity were to pass through, the data loggers would capture it and record its exact movement.

Magnetometer: An instrument that measures the strength of the magnetic field in a given area.

Object manipulation: Any inanimate object moving apparently of its own accord. If gravity, air current, and other normal explanations can be eliminated, paranormal investigators believe that a ghost may have created the motion by harnessing energy and directing it at the object in question.

Orb: A ball or circle of light often found in videos or photographs. Some people believe that an orb is the energy or the spirit of an individual who has passed, though orbs are generally simply reflections or bits of dust.

Parabolic microphone: A highly sensitive microphone that uses a parabolic reflector (satellite dish) to pick up and direct sound waves onto a receiver. Often used in law enforcement and espionage work. A paranormal investigator listens to sounds using large headphones and redirects the parabolic microphone to better pick up unusual noises as they occur, allowing for more real-time data analysis.

Paranormal: Anything outside of the normal range of human experience and scientific explanation.

Personal experience: A sensory encounter with a ghost or another paranormal phenomenon by a living human. Personal experiences can be visual (seeing an apparition), auditory (hearing a disembodied voice), olfactory (becoming aware of an unusual odor), or tactile (getting pushed by an invisible force, feeling short of breath, etc.).

Poltergeist: A noisy spirit. The term comes from the German words *poltern* (to make noise) and *geist* (ghost).

Provoking: Attempting to incite paranormal activity with loud verbal commentary or taunts.

GLOSSARY

Real time: As an event unfolds.

Renovation Theory: Holds that remodeling an old building can change its energy level and agitate spirits inside it, kickstarting bursts of paranormal activity.

Residual haunting: A paranormal event or entity that shows no awareness of the living world and does not interact or communicate. Residual hauntings generally involve "playing back" a significant event or a routine from an entity's life.

RT-EVP recorder: Supersensitive digital audio recorders that contain two microprocessors—one that records and allows an investigator to hear EVPs (or ghost voices) in real time (RT) and one with a programmable four-second recording delay or other interval.

Saturation: The first wave of a paranormal investigation, this involves taking photographs and establishing baseline readings of environmental levels in the absence of perceivable paranormal activity.

Scientific method: A technique used to investigate a subject or field of study. It involves forming a hypothesis; collecting measurable, empirical data to test the hypothesis through observation and experimentation; analyzing data; refining hypothesis if necessary.

Shadow people: Shadow people are dark shapes with an outline that resembles a human body but without distinguishable facial features or other detail. They can appear in photographs or on video, though they may be invisible to the naked eye. It is thought that these spirits do not have enough energy to materialize as full-bodied apparitions. We believe these are intelligent manifestations.

Shotgun sweep: Involves placing digital data loggers throughout a specific location (like shotgun pellets) to record changes in the environment that may indicate the presence of paranormal activity.

Thermal-imaging camera (or infrared camera): A camera that measures the heat an object radiates and takes readings of temperature ranges.

VP: Voice phenomena that are within the range of audible human perception. These phenomena are believed to represent an even higher level of energy than EVP.

Walk-through: One of the first steps in any Everyday Paranormal investigation, this involves examining the entire site to be investigated (buildings and grounds) on foot during daylight hours with particular focus on areas reported to have experienced paranormal activity.

Bibliography

Chapter 2: The Phantom Regiment
Official Web site for the Jennie Wade House, Gettysburg Tour Center online. http://www.gettysburgbattlefieldtours.com/jennie-wade-house.php

Chapter 3: Tourists in Ghost Land
Interview with Docia Williams

Lyons, Linda, One-Third of Americans Believe Dearly May Not Have Departed," Gallup, July 12, 2005. http://www.gallup.com.

Official Web site of the Menger Hotel. http://mengerhotel.com/page/ntm5/The_Menger_History.html.

Moore, David W., "Three in Four Americans Believe in Paranormal," Gallup, June 16, 2005. http://www.gallup.com.

Chapter 4: Voices from Cell Ten
"Alcatraz." http://www.alcatrazhistory.com/

Alcatraz Island: An Inescapable Experience. http://www.alcatrazcruises.com.

Official Web site of the National Park Service, "Alcatraz Island." http://www.nps.gov/alca/index.htm

Novak, Laura. "Flocking to See Alcatraz," *New York Times*, March 26, 2006. http://travel.nytimes.com/travel

Shermer, Michael, "Patternicity." *Scientific American*, December 2008. http://www.michaelshermer.com/2008/12/patternicity/

Chapter 5: To Catch a Ghost
Official Web site for Fort Sam Houston. http://www.fortsamhoustonmwr.com.

San Antonio Convention & Visitors Bureau. http://www.visitsanantonio.com.

Chapter 6: Chilling Evidence from an Ice Rink
Interviews with San Antonio Conservation Society staff members.

Chapter 7: Echoes in an Abandoned Old Folks' Home
"Nursing Home Abuse Statistics." http://www.nursing-home-abuse-resource.com/care_center/nursing_home_statistics.html.

"Nursing Home Abuse Statistics," March 4, 2009. http://www.articlesbase.com/health-and-safety-articles/nursing-home-abuse-statistics-801584.html.

Nursing Home Statistics (AHCA). http://www.efmoody.com/longterm/nursingstatistics.html.

Chapter 8: Mysteries at Myrtles
Moses, Jennifer, "More Than Just a Place to Stay: Louisiana; Ghosts and atmosphere galore in a restored plantation house," *New York Times,* July 25, 1999.

Official Web site of the Myrtles Plantation. http://www.myrtlesplantation.com/

Chapter 9: Ghosts of the Garden District
Dexter, Ralph W. "The Early Career of John L. Riddell as a Science Lecturer in the 19th Century," *Ohio Journal of Science,* vol. 88, 1988, pp. 186–188.

Official Web site of Gainesville Coins, "New Orleans U.S. Mint History." http://www.gainesvillecoins.com/New-Orleans.aspx

Official Web site of the Louisiana State Museum, the Cabildo, "Antebellum Louisiana: Disease, Death, and Mourning." http://lsm.crt.state.la.us/cabildo/cab8a.htm

Marland, Hilary. "The Medical Activities of Mid-Nineteenth-Century Chemists and Druggists, with Special Reference to Wakefield and Huddersfield," *Medical History*, 1987, 31: 415–439.

Official Web site of Tulane University, history. http://www.tulane.edu/~mrbc/2001/MRB%20WEBPAGE/history.htm

Chapter 10: Whispers in the Walls

Official Web site of the Texas Parks and Wildlife Department, "Enchanted Rock State Natural Area." http://www.tpwd.state.tx.us/spdest/findadest/parks/enchanted_rock/

Official Web site of the Texas State Historical Association, "Enchanted Rock Legends." http://www.tshaonline.org/handbook/online/articles/lxe01

Official Web site of the United States Geological Survey, "The Geology of Radon." http://energy.cr.usgs.gov/radon/georadon/3.html.

Chapter 11: The Flying Tape Fiasco
IEEE Global History Network, "WAOI San Antonio Texas." http://www.ieeeghn.org/wiki/index.php/WOAI_San_Antonio_Texas.

Interviews with San Antonio Conservation Society staff members.

"KMOL-TV May Not Get Designation," *Express-News*, November 9, 1988.

Chapter 13: Ancient Artifacts and Disembodied Voices
Official Web site of the Caddo Nation. http://www.caddonation-nsn.gov/

Goldman, Kay, "Review of *The French in Texas: History, Migration, Culture* by François Lagarde (University of Texas Press, 2003)," *H-France Review*, Society for French Historical Studies, vol. 4 (March 2004), no. 31.

Official Web site of the Institute of Texan Cultures. http://texancultures.com/

"Texas Indians." http://www.texasindians.com/

Oracle ThinkQuest Education Foundation. http://library.thinkquest.org/J001272F/history/french.htm

BIBLIOGRAPHY

Powell, Kimberly, "Death & Burial Customs: Traditions & Superstitions Related to Death," http://genealogy.about.com/od/cemetery_records/a/burial_customs.htm

Official Web site of the U.S. Department of the Interior, National Park Service, "Burial Customs and Cemeteries in American History." http://www.nps.gov/history/nr/publications/bulletins/nrb41/nrb41_5.htm

Chapter 14: Library Specters

"The Dienger Building." http://voicesofthetexashills.org/vthhbldg0002.htm.

Official Web site of Boerne, Texas, "Library History." http://www.ci.boerne.tx.us/index.aspx?NID=205.

Chapter 16: Sounds from Beyond

Clarkson, Michael. *Poltergeists: Examining Mysteries of the Paranormal.* Toronto: Firefly Books, 2006.

Wagner, Stephen. "The Bell Witch: The tormenting spirit of America's best-known poltergeist case," http://paranormal.about.com/od/trueghoststories/a/aa041706.htm

"Wartburg Castle." http://www.sacred-destinations.com/germany/wartburg-castle

"Wartburg Castle: Martin Luther, His Bible, His Devil and a Saint." http://www.bargaintraveleurope.com/08/Germany_Wartburg_Castle_Eisenach.htm

Chapter 17: School for Scares

Official Web site of Wiederstein Elementary School, "Who Was Mr. O. G. Wiederstein?" http://www.scuc.txed.net/OGWiederstein.cfm?subpage=8874.

Chapter 18: The Battle over *Ghost Lab*

"Haunted Black Swan Inn." http://www.informationsanantonio.com/toseeanddo/historicsanantonio/sanantonioghostsandlegends/ghostsofblackswaninn.html.

Official Web site of the Texas State Historical Association, "Battle of Salado Creek." http://www.tshaonline.org/handbook/online/articles/qfs01

Official Web site of Victoria's Black Swan Inn. http://www.victoriasblackswaninn.net/

Made in the USA
Middletown, DE
07 January 2018